Cambridge Studies in Biological and Evolutionary Anthropology 24

Migration and colonization in human microevolution

Migration and colonization are major forces affecting the frequency, spatial pattern and spread of genes in human populations. Here, Alan Fix reviews theories of migration developed by biologists and social scientists, and surveys patterns of migration in a diverse sample of human populations. Using these empirical studies, he evaluates models of migration developed by population geneticists and explores more realistic models using computer simulation. He then shows the relevance of studies of migration as a microevolutionary process to the understanding of longer term global patterns of human diversification, by examining the spread of anatomically modern *Homo sapiens*, the demic diffusion of agriculture in Europe, and the origins of human diversity in the Malayan Peninsula. By focusing on migration as a process rather than as its genetic consequences, the book provides a bridge between biological and social science studies of migration, genetic microevolutionary theory and longer term human evolution.

ALAN G. FIX is Professor of Anthropology at the University of California, Riverside. His primary research interest involves the population genetics and demography of small-scale human populations, work which is based on extensive fieldwork with the Semai Senoi of Malaysia.

T0275649

Cambridge Studies in Biological and Evolutionary Anthropology

Selected titles also in the series

11 *Genetic Variation and Human Disease* Kenneth M. Weiss
    0 521 33660 0 (paperback)
12 *Primate Behaviour* Duane Quiatt & Vernon Reynolds
    0 521 49832 5 (paperback)
13 *Research Strategies in Biological Anthropology* Gabriel W. Lasker & C.G.N.
    Mascie-Taylor (eds) 0 521 43188 3
14 *Anthropology* Stanley J. Ulijaszek & C.G.N. Mascie-Taylor (eds)
    0 521 41709 8
15 *Human Variability and Plasticity* C.G.N. Mascie-Taylor & Barry Bogin (eds)
    0 521 45399 2
16 *Human Energetics in Biological Anthropology* Stanley J. Ulijaszek
    0 521 43295 2
17 *Health Consequences of 'Modernisation'* Roy J. Shephard & Anders Rode
    0 521 47401 9
18 *The Evolution of Modern Human Diversity* Marta M. Lahr 0 521 47393 4
19 *Variability in Human Fertility* Lyliane Rosetta & C.G.N. Mascie-Taylor (eds)
    0 521 49569 5
20 *Anthropology of Modern Human Teeth* G. Richard Scott & Christy G. Turner II
    0 521 45508 1
21 *Bioarchaeology* Clark S. Larsen 0 521 49641 1 (hardback),
    0 521 65834 9 (paperback)
22 *Comparative Primate Socioecology* P.C. Lee (ed.) 0 521 59336 0
23 *Patterns of Human Growth, second edition* Barry Bogin
    0 521 56438 7 (paperback)

Migration and colonization in human microevolution

ALAN G. FIX
University of California, Riverside

CAMBRIDGE UNIVERSITY PRESS
Cambridge, New York, Melbourne, Madrid, Cape Town, Singapore, São Paulo

Cambridge University Press
The Edinburgh Building, Cambridge CB2 2RU, UK

Published in the United States of America by Cambridge University Press, New York

www.cambridge.org
Information on this title: www.cambridge.org/9780521592062

First published 1999
This digitally printed first paperback version 2005

A catalogue record for this publication is available from the British Library

Library of Congress Cataloguing in Publication data

Fix, Alan G.
Migration and colonization in human microevolution / Alan G. Fix.
   p.   cm. – (Cambridge studies in biological and evolutionary anthropology)
Includes bibliographical references and index.
ISBN 0 521 59206 2
1. Human population genetics.   2. Emigration and immigration.
3. Human evolution.   I. Title.   II. Series.
Gn289.F59   1999
304.8 dc21   98-32340   CIP

ISBN-13 978-0-521-59206-2 hardback
ISBN-10 0-521-59206-2 hardback

ISBN-13 978-0-521-01954-5 paperback
ISBN-10 0-521-01954-0 paperback

To Betsy, Aaron, and Amy

Contents

Preface

This book is about the role of migration and colonization as agents in human microevolution; that is, how human spatial displacements affect the frequencies, pattern, and spread of genetic variants in human populations. My approach is from anthropological genetics: anthropological in that patterns of mating, technologies and economies, and social structures are cultural attributes of populations that profoundly influence migration; genetic in that it is the biologically significant effects of migration on genetic structure that are to be understood.

The study of migration ranges across the territory of several disciplines, from biology and evolutionary genetics to the social sciences, including cultural anthropology, archaeology, demography, and geography.

In biology, migration (or gene flow) is one of the four forces of evolution along with mutation, natural selection, and genetic drift, and plays an important role in the general theory of evolutionary genetics (Wright 1931). Migration is also a major topic in ecology both as a phenomenon to be understood (Dingle 1996) and as a variable in the relationship between social and genetic structures (Chepko-Sade & Halpin 1987).

In the social sciences, geographers, with their focus on spatial relationships, have been at the forefront of migration research (Clark 1986; Lewis 1982; Robinson 1996). Migration has always been an important explanatory concept for archaeologists. Recently this interest has intensified (Anthony 1990) and colonization has been raised to the central issue of human prehistory by Gamble (1994) in his book, *Timewalkers: the Prehistory of Human Colonization*. Such explanations have a venerable history in cultural anthropology as well (recall the various 'historical schools' of anthropology active in the early part of the century (Lowie 1937)).

In human genetics, the rapid proliferation of molecular 'markers' in recent years has stimulated efforts to trace routes of migration and colonization. Global questions of human ancestry from the initial spread of our species (Cann *et al.* 1987) to new hypotheses for the peopling of the New World (Wallace & Torroni 1992) and the colonization of Oceania (Serjeantson & Hill 1989) and the demic diffusion of agriculture through

Europe (Ammerman & Cavalli-Sforza 1984) have been addressed using both molecular and 'classical' genetic data. Perhaps the most ambitious of these studies has been the global survey of gene distributions by Cavalli-Sforza and his colleagues (1994); the predominant explanation offered for these patterns in this work is population fission, migration, and isolation.

The fundamental problem for all reconstructions that assume migration as a cause, however, is how to distinguish past population movements from other processes that lead to the same distributional outcomes. More particularly, how do present distributions of artifacts, languages, or gene 'markers' signify movement of people as opposed to other processes such as cultural diffusion or transformation (in the case of genes or natural selection)? For this reason, the definitive demonstration of any prehistoric movement must depend on multiple converging lines of evidence.

Critical to this demonstration must surely be the understanding of the *process* of movement (Madsen & Rhode 1994). The comparative study of migration in the contemporary world is the primary source for a causal theory of migration. This is not to say that conditions in the distant past could not have been so different that no modern analog would apply; however, a *theory* of migration specifying the variables and conditions affecting rates and patterns of migration should provide guidance for assessing models of past population movement. The primary sources for this theory are the large literature from anthropological genetics documenting comparative population structures and mobilities (e.g., Boyce 1984; Crawford & Mielke 1982; Harrison & Boyce 1972b; Little & Leslie 1993; Mascie-Taylor & Boyce 1988) and the ethnographic record documenting covariation between socio-ecological variables, population properties, and movement (Johnson & Earle 1987).

My aim in this book is to survey this literature, looking for patterns and regularities among the diversity of cultures and population structures recorded by anthropologists, and attempting to identify key variables affecting gene distributions. A general theory of migration must be able to encompass the full range of human mobilities, from the extreme local sedentism found in some small-scale agrarian societies to the mass migrations of recent times. Such a comprehensive theory would require enlarging of the scope of classic genetic models to include a greater diversity of population structures. Simple patterns of nearest neighbor exchange or continent to island gene flow capture only a portion of this diversity. More complex models, such as those made possible by computer simulation, allow experimentation with a greater range of population structures and migration patterns. While I cannot claim to have developed a general model in this book, I have tried to suggest ways to broaden our view of the role of migration and population structure in human microevolution.

The title of a paper by John Moore (1994), 'Putting anthropology back together again', nicely captures my orientation in this work. For much of its history, anthropology constituted a kind of 'natural historical' study, integrating data and questions from biology to art. Current trends seem to be toward more specialization and fragmentation of the subdisciplines. A microscope gives a very clear picture of a very small field. If we are still to be concerned with global questions of human diversity and evolution, too narrow a focus blinds us to this broader picture. To know everything about potsherds (or the mitochondrial genome) and little about people does not seem a good strategy to understand human populations. Such understanding is more likely with the exchange of information across disciplinary (or subdisciplinary) lines along with efforts to synthesize these data and interpretations.

The dilemma of synthesis is to venture into fields where one is not expert on the one hand, and to fail to communicate to experts in fields other than one's own on the other. I hope to have avoided major mistakes in areas where I have little technical expertise. I have also tried to avoid jargon and excessive equations. Although the book is primarily intended for anthropologists and human geneticists, I would also be gratified if biologists interested in broad patterns across taxa (Dingle 1996) might be made aware of the scope of detailed data on migration in human populations from anthropological and historical sources.

Plan of the book

The first chapter surveys the perspectives of the different disciplines concerned with the study of migration and provides definitions of basic terms and an introduction to causal models. This discussion provides a general context for the anthropological and genetic study of migration in the remainder of the book.

Human populations throughout history have varied enormously in size, density, and structure, reflecting their diverse cultural and ecological circumstances. Chapter 2 begins with a sketch of the techno-economic variation among human populations and its implications for population densities and migratory patterns. The principal factor implicated in this diversity is the degree of intensity of land use required by the technology of the population which is also correlated with population density and degree of social integration. Using these variables as dimensions of comparison, a series of anthropologically well-known case studies are presented. These represent points on a continuum of populations from low-density, highly mobile, politically autonomous families or local groups, through swidden-farming village-level polities, to more densely settled farmers organized into regional political entities and include:

(1) Classic mobile hunter–gatherers showing flexible group structure and wide ranging kin ties resulting in large interacting population networks. Examples discussed in Chapter 2 are Australian Aboriginal groups (mainly the Yolgnu of Arnhem Land), the !Kung San of the South African Kalihari Desert, and the central African forest Aka pygmies.

(2) More sedentary collectors with higher population densities and more localized groups. Examples of this category would include the Northwest Coast-California Indians (and maybe Mesolithic Europeans). Unfortunately, this group of societies has not received the detailed study devoted to the more mobile foragers, and therefore no examples have been included. Because most of these societies occupied lands that were attractive to farmers, few of them persisted to be documented by anthropologists.

(3) Small scale agricultural societies with population densities no higher than the settled collectors of category 2 and practicing extensive land use techniques (swiddens). Politics is local in these societies; villages are usually autonomous and subject to periodic group fission and fusion. Examples considered in Chapter 2 are Amerindian groups of the Vaupés region of Columbia and the Venezuelan Yanomamo and the technologically similar societies of the Malaysian Semai Senoi and the Gainj of Papua New Guinea.

(4) More intensive agriculturists with higher population densities, larger political groups including state organization and social stratification. The societies representing this category discussed in Chapter 2 are all examples of what are often called 'peasants'; that is, rural agrarian peoples within the borders of nations. They include the Basque farmer/herders of the French–Spanish border region, rural parishes in Oxfordshire, England and a subdistrict in Uttar Pradesh, northern India.

For each case, the factors affecting migration will be examined and special issues illustrated by the case will be detailed (e.g., the role of inheritance patterns in Basque migration). The rationale for considering these different cultures and economies is to assess the extent of variation in human migratory patterns and its relevance to genetic models.

The aim in Chapter 3 is to characterize the classic population genetics models of migration and population structure. These include the island, isolation by distance, stepping-stone, migration matrix, and neighborhood knowledge models. The survey is not exhaustive nor very mathematical (see Jorde, 1980, for a detailed review), but rather focuses on the underlying migration patterns that these models are attempting to capture. The strengths and weaknesses of these models are evaluated in light of the previous discussion of the pattern and structure of human migration and, for several, case studies are examined as focused examples. A set of basic variables is identified and considered, including: (1) stage of the life cycle at which migration occurs, whether pre-marital, marital, or post-marital; (2) the units of migration, whether gametes or individuals along a continuum to population fission and group migration; (3) the structure of migrant groups (random or kin); (4) the spatial pattern of migration varying from exchange between nearest neighbors along a continuum to long distance displacements; (5) geographic structure and distance; and (6) population size.

Chapter 4 continues the discussion of population genetics and migration now focused on computer simulation modeling, a method allowing a more detailed exploration of some of the variables identified in Chapter 3 (for example, the role of kin-structured migration was first studied using simulation techniques). Evolution, as Sewall Wright long maintained, is an interaction of forces; gene frequency change is often the joint outcome of several processes operating concurrently. Particularly in the small populations studied by anthropologists, these processes are subject to random variation. Computer simulation allows these more complex interactions and stochastic variation to be modeled. In a broader sense, simulation can be viewed as an 'experimental' technique to study history. In the absence of time machines, we can't turn back the clock to see what happened in evolution. We can, however, replicate processes under controlled conditions using computer simulation and see what might have happened.

Several examples of stochastic simulation models incorporating the interaction of migration with other forces of evolution are presented, including the effect of kin-structuring of migration (KSM), the interaction of clinal and balanced selection with KSM, the wave of advance of an adaptive allele, the potential role of colonization and founder effect in producing genetic clines, and the interaction of colonization and extinction in metapopulations. The role of stochastic variation in demography and marriage pools (and the consequent need for outmarriage in small populations) is also explored in a simulation model.

Chapter 5 considers the implications of contemporary human migration studies for understanding and reconstructing population movements over the longer span of prehistory and human evolution. A major problem is reconciling the range expansions of humans into previously unoccupied regions (the Americas, remote Oceania) with the ethnographic record of mostly local movements. Three examples of suggested large scale colonizations or invasions are presented: the postulated initial spread of *Homo sapiens* from Africa; arguments for and against the spread of agriculture through Europe by Near Eastern migrants; and traditional migratory wave explanations for human diversity in the Malayan Peninsula. All three of these cases demonstrate the great difficulty of discriminating among migration and other causes of genetic distributions. This inherent multi-causality reinforces the need for a better understanding of the process of migration and for a general evolutionary theory of its causes and consequences.

Acknowledgments

I would like to thank Gabriel Lasker for suggesting that I write this book and Derek Roberts for his encouraging words on the proposal. Thanks also to Tracey Sanderson at Cambridge University Press for providing further encouragement and helpful advice.

My enormous intellectual debt to the many anthropologists and geneticists who have thought and published on migration will be obvious in the pages that follow. My interpretation of this literature was shaped by my major professor, Frank Livingstone, whose study of the interaction of genetics, ecology, and culture (Livingstone 1958) has always epitomized to me the anthropological approach to human evolution.

The experience of living for over a year with Semai people in Malaysia had a profound effect on my thinking. The adage that fieldwork is crucial to the development of an anthropologist may be old but it is true. I am very grateful to the many Semais who were my teachers in Malaysia. Research in Malaysia was supported by the National Institute of Mental Health and the Wenner Gren Foundation for Anthropological Research.

Sabbatical leave time to complete parts of the text was given by the University of California, Riverside; the Academic Senate of the University also provided financial support through Intramural Research grants.

I would also like to thank Dee Baer, who read and made helpful suggestions on portions of the text, Bryan Epperson for illuminating discussions of spatial modeling in genetics, and Susan Mazur, who provided assistance with the preparation of the manuscript, and Karl Taube, who suggested (and redrew) the cover image of the four god-bearers leaving Aztlan on their epic migration to Tenochtitlán.

1   *The study of migration*

Evolutionary and ecological perspectives

Definitions and patterns: migration, dispersal, and gene flow

Migration is a process with consequences that are important to theory in several disciplines and the term may be used in several senses in these different domains. Biologists study migration both from an ecological and an evolutionary point of view. That is, migration takes organisms into different habitats thereby affecting resource availability and other ecological parameters; likewise, migration may result in gene flow with effects on the gene pool of recipient populations.

The spectacular long-distance to and fro trips of birds and butterflies are more likely to be called 'migration' in biology than one-time, non-return movements that are termed 'dispersal' (Dingle 1996). Both are ecologically similar in that they are usually efforts to improve environmental conditions for the organism and, for Dingle (1996) at least, both types of movements are behaviorally alike in that they involve locomotion not seen at other times in the life cycle. In fact, Dingle would like to reserve the definition of migration for the characteristic pattern of behavior rather than the outcomes of movement (dispersal or aggregation). Moreover, he would exclude accidental or unintentional movement from the definition since these are not behaviors potentially subject to natural selection. Thus migration is 'persistent and straightened-out movement effected by the animal's own locomotory exertions' (Dingle 1996:25) carrying them to new habitats. This behavior will have ecological effects that will determine its evolution.

Evolutionary geneticists on the other hand generally use 'migration' as a synonym for 'gene flow' (Merrell 1981); that is, one-way movement to a new population. Strictly speaking, the migration coefficient in population genetics, m, is not the number of individuals moving (the quantity of interest to population biologists studying demographic change) but rather a measure of the proportion of gametes contributed by immigrants to the gametic pool making up the next generation in the recipient mendelian population.

1

This usage is in sharp contrast to Dingle's (1996) behavioral definition since it is an (highly abstract) *outcome* that is the relevant factor.

Population structure: units of analysis

The study of migration presumes clearly defined populations with migrants moving between them. In demography, populations are generally administrative units ranging from nations (international migration) through various state, county, or metropolitan polities (internal migration). Geneticists and anthropologists define the population unit by focusing on the relevant behaviors linking together the members of the group; interbreeding in the case of genetics, shared language, culture and social interaction for anthropology. Just as there is a hierarchy of political units, social or breeding populations may vary from local groups to ever larger partitions of the species. In practice, then, migration may be measured and modeled between villages, parishes, counties, districts or countries; all are valid units depending on the problem being investigated (Fix 1979).

Causal models

At the most general, migration is movement to a new location undertaken to improve the environmental conditions for the organism (Dingle 1996). However, simply foraging in search of widely scattered food does not satisfy Dingle's (1996:54) behavioral definition of migration since it does not entail focused movement to a new habitat. Thus organisms in patchy environments might be expected to be nomadic, tracking seasonally shifting or widely dispersed resources. This mobility, while technically not migration in Dingle's sense, may nonetheless involve considerable distances. For instance, Lee (1980) estimated that !Kung women in the Kalahari travel some 2400 km per year, mostly in day-trips of less than 10 km, in the food quest. Many human foragers provide excellent examples of nomadic movement over large ranges exploiting highly localized, often high caloric-return food items. Indeed, this life way has been suggested as part of a key adaptive shift in hominid evolution (Kurland & Beckerman 1985).

Nomadism within a home range may be an appropriate response to patchy resources; however, when the habitat deteriorates or is intrinsically ephemeral, migration to another range may be forced (Dingle 1996:270). In the case of many bird species, long range migration occurs to avoid seasonal shortages (brought on by winter) and to take advantage of short-term abundance (e.g., as occurs during the arctic summer). Temporary abundance may not be only seasonal. For instance, Australian banded stilts inhabit marshes and salt lakes (Dingle 1996:55). Only after heavy rains fill normally dry lake beds do these birds migrate in great numbers (up to 100,000 birds) to breed. Apparently the temporary abundance of brine shrimp to feed the young birds triggers migration and breeding. Such events may occur at intervals of several years in the arid regions of South Australia.

Bird migrations are often round trips, especially seasonal movements from summer to winter range and return. Such periodicity implies relatively stable or predictable conditions in both ranges. Movement oscillates between two known habitats. One-way migrations, however, are more likely in uncertain or unpredictable environments (Dingle 1996:61). When the environment deteriorates with no indication of when it will improve again, it may be abandoned for a new range.

In theory, it is possible to calculate the costs and benefits of migration just as for any other behavior. These combined with the constraints of the environment, should predict the occurrence and frequency of migration. Migration should evolve as a function of the cost of migration, the availability of alternative habitats, and the basic ecological parameters of population growth rate, r, and carrying capacity, K (Dingle 1996:271). This theory has been presented as alternative life histories under the control of different selective regimes called 'r and K' (Pianka 1970; see Stearns, 1992 for a critique). R species inhabit temporary habitats that put a premium on high intrinsic rates of growth, r. Production of large numbers of quickly maturing young ensures that these offspring can disperse widely in search of ephemeral habitats. K species tend to occupy permanent habitats with high carrying capacities. Selection under these conditions is for fewer offspring with greater competitive ability rather than large numbers of highly motile progeny. Clearly this is not an absolute dichotomy – species may possess r-like attributes while at the same time be selected along the K dimension. For example, humans have many attributes of a quintessential K species: long life span; relatively few offspring with heavy parental investment; etc. Nonetheless, compared to our closest relatives, the African great apes, we show some attributes that are more r-like (Lovejoy 1981), especially a shortened birth interval. Thus species or populations exploiting patchy, uncertain environments might be selected for superior colonizing ability (r selection) *relative* to related species or populations in more stable habitats in which K selection might be more important. This point is particularly relevant to assessing the ability of humans to rapidly colonize large areas such as the Americas and Australia (see Chapter 5).

The general conclusion to be made is that migration should be more important where habitats are temporally transient or spatially patchy. The greater the degree of environmental uncertainty, the greater the potential strategic importance of migration.

Caloric considerations are only part of the evolutionary equation, however. Mating and reproduction (genetic effects) must be accomplished if individuals are to leave offspring (and their genes) in subsequent generations. Finding a mate may be an important cause of migration and, indeed, marriage is often the primary determinant of movement from the birthplace in many sedentary human societies.

Of course, marital movement also may have somatic consequences. Tylor's (1888:267) famous aphorism, 'marry out, or be killed out' identifies the political advantages of exogamy. Wide-ranging marital ties extend affinal kinship networks potentially reducing conflict and often allowing access to resources in time of need. Such marital systems might be expected to occur in the same contexts that non-marital migration would be favored; that is, where resources vary greatly in time and space (see Chapter 2 for some examples).

Costs and benefits relating to mating dispersal can be measured in reproductive as well as caloric currencies. Shields (1987) provides an extensive discussion of the fitness effects of philopatry (non-dispersal from natal site) versus dispersal. Table 1.1 is a synopsis of some of these key factors derived from his Table 1.3 (1987:15–16).

A time honored argument for the direct genetic benefit of out-mating is the avoidance of inbreeding depression (for humans, see Aberle *et al.* 1963). Increased homozygosity of rare deleterious recessive alleles leading to the phenotypic expression of genetic diseases would seem an obvious disadvantage of endogamy (or philopatry), in so far as it increased mating with relatives. However seemingly obvious, the actual degree of debility caused by inbreeding in humans has never been satisfactorily documented (Bittles & Makov 1988). In this regard, Shields (1987:19) cites data from acorn woodpeckers showing that some 20 percent of groups were closely inbred, usually within nuclear families. In the same volume (Chepko-Sade & Halpin 1987), a number of studies of other species that also practice high levels of inbreeding are described. Closer to home for anthropologists, the ubiquity of male dispersal in cercopithecine monkeys has been explained as avoidance of incest and thereby inbreeding depression (Bischof 1975). However, other competing hypotheses have not been excluded even in this well known case (Shields 1987).

The other side of the genetic coin from inbreeding depression is the potential cost associated with outbreeding (Shields 1987). When organisms disperse, they may enter environments with different selective conditions than those to which they are adapted. Problems faced by human dispersers

Table 1.1. *Costs and benefits of mating dispersal*

Potential benefits
I. Genetic
   A. Avoid inbreeding depression
II. Somatic
   A. Direct fitness benefits
      (1) Escape local crowding; gain access to resources including mate
   B. Indirect fitness benefits
      (1) Avoid competition with sedentary kin

Potential costs
I. Genetic
   A. Outbreeding depression
      (1) Disrupt potentially coadapted gene complexes
   B. Migration load
      (2) Enter environment with genotype not locally adapted to disease or other
         factors
II. Somatic
   A. Direct fitness effects
      (1) Risk and energy expenditure of migrating
      a. Lack of familiarity with new locale reducing foraging efficiency
      b. Energy cost of migrating and increased exposure to predators
      c. Greater susceptibility to local diseases
   B. Indirect fitness effects
      (1) Lack of mutual aid and support for non-dispersing kin

in the island of New Guinea illustrate this situation. Malaria in New Guinea differs greatly in incidence from location to location (Bayliss-Smith 1994). On account of this variability, Bayliss-Smith (1994:305) argues that 'exogamous marriage is seen as a particularly risky practice in malarial areas'. Where genetic resistance to endemic malaria has evolved, movement bears fitness costs. Strain specific immunity acquired in childhood would not equip a migrant to resist a different strain in the new environment. Similar arguments can be made for any patchily distributed disease – brides or grooms moving to a new area might encounter diseases or strains to which they had no evolved or acquired immunity.

A more tenuous potential fitness cost of dispersal depends on the presence of local coadapted gene complexes. This concept has long been championed by Ernst Mayr (1963) but actual evidence for such sets of interacting genes promoting adaptation to local environments is scarce. Where such complexes exist, migration between groups might disrupt the favorable gene combinations and therefore reduce progeny fitness.

Somatic costs and benefits of dispersal are also summarized in Table 1.1. The general sense of migration as a strategy to improve resource acquisition presented by Dingle (1996) certainly is a primary somatic factor both in the direct fitness value gained by the disperser or indirectly by no longer competing for local resources with kin remaining in the natal group. On the negative side of the ledger, the energy and risk associated with migration may be high depending on the distance, possible barriers to travel, and risk factors such as predators (including disease micropredators). Where knowledge of local conditions are important for foraging, unfamiliarity reduces efficiency. Particularly for social species such as humans, dispersers may directly suffer due to lack of mutual support provided by kin in the new locale and indirectly due to inability to support kin remaining at home.

A strictly genetic accounting weighs the costs of inbreeding against the costs of breaking up adaptive gene complexes and/or maladaptation to the new environment. But genetic factors can not be considered in isolation from the fitness effects due to somatic factors. Indeed, there is no *a priori* necessity that any one factor, genetic or somatic, will predominate in all cases. While inbreeding has often been seen as the critical problem to be resolved by dispersal, in some species and/or environments other factors may be as or more important. Shields (1987) points out that Bengtsson's (1978) assumption that the genetic costs of philopatry are balanced by the greater somatic costs of dispersal could just as easily be turned around. The cost of philopatry could be increased competition within the natal population (a 'somatic' factor) balanced by the 'genetic' cost of outbreeding depression. Some species tolerate apparent high levels of inbreeding without obvious genetic deterioration; others seemingly accept the high somatic costs of dispersal. Theory can identify the relevant variables but the values taken by each variable may vary with different empirical situations. The diversity of possible outcomes in mammalian dispersal (literally from mice to humans) and evaluations of the causal factors can be found in Chepko-Sade & Halpin (1987).

Social science perspectives

Human migration is the concern of demographers, geographers, anthropologists, sociologists, and economists. All of these disciplines share overlapping interests and concepts; however, particular emphases differ among the fields.

Demographic models

For demography, migration is one component of the basic demographic equation (Newell 1988:8):

$$P_{t+1} = P_t + B - D + \text{Inmigrants} - \text{Outmigrants}$$

where P_{t+1} is the population after one unit of time which is dependent on the initial population at time $t(P_t)$, the numbers of births (B) and deaths (D) occurring between time t and $t + 1$, and the number of in- and outmigrants to the population during the same time interval. The difference between the number of births and deaths is referred to as 'natural increase' and 'net migration' is the corresponding migrational differential.

P, 'the population', can be any size unit of interest although usually demographers study nations since these governments provide the statistical data most available to them. The definition of migration in this equation depends on the population referent. Births and deaths occur *within* that unit and migration is something that happens *between* those units. Because of the focus on nation-states, immigration and emigration are defined strictly in international terms while the terms 'inmigration' and 'outmigration' are applied to internal movement (Newell 1988). Anthropologists (including anthropological geneticists), in contrast to demographers, are much more likely to be interested in local populations. Nonetheless, many of the same problems of definition and measurement occur at the local level as exist for nations.

The first point to notice about the demographic equation is that only the *number* of individuals migrating in and out of the population are specified. The *structure* of migration, either in terms of the usual demographic markers of age and gender or in terms of spatial location of migrants, is not considered. Elaborate methods for more precisely characterizing births and deaths including age specific fertility rates and life tables have been devised by demographers. Migration, however, has not received such sophisticated treatment.

One reason for this relative neglect of migration by demographers is the intrinsic difficulty of measuring migration. Part of the problem is that of definition. Births are discrete events that occur to women of definable age. Likewise, deaths occur only once to everyone. Both events are recorded in national registries that provide data to demographers. In contrast, migration is less clearly marked, may occur repeatedly, may be reversed (return migration), and therefore is much harder to measure. National migration statistics may be available for some countries but they are not of the same degree of precision as birth and death statistics.

Consider, for example, the conventional definition of an immigrant by a demographer (Newell 1988:84): 'a person who has resided abroad for a year or more and, on entering the country, has declared an intention to stay for a year or more'. Clearly, the arbitrary unit of time and the inference of intention distinguish migration from the 'natural' events, birth and death. Similarly, for internal migration, a problem arises when the distance of movement is short. Clark (1986:12) reserves the term 'migration' for relatively permanent moves that are 'too far' for continued commuting. Recent trends in Southern California have extended this distance to a previously unimaginable degree with people routinely traveling 40 to 60 miles (64–96 km) one-way daily to work. Clark (1986) also notes the problem of using governmental units to define the scale of migration. 'Internal' migration within the boundaries of the United States, for example, may encompass 2000 miles (3218 km), a distance that would cross many national boundaries in other parts of the world.

These spatial concerns extend the study of migration beyond the disciplinary limits of traditional demography into that of human geography (Clark 1986; Lewis 1982; Robinson 1996). Similarly, the economic and cultural constraints on and consequences of migration make it an important topic of study for economists and anthropologists.

Geographic and economic models

The 'classic questions' defining the domain of study for social scientists are: 'who moves, why do they move, where do they move, and what are the impacts when they get there' (Clark 1986:10).

'Laws' of migration

Historically, models of human migration have been dominated by economic variables. For most of these classic models, movement occurred as individuals were pulled by economic forces to destinations offering better opportunities perhaps having been pushed (also by economic factors) from their home locales. Job seeking thus was the principal motivator for movement. Migration achieved spatial equilibrium in income and employment. Flows of migration were from areas of low wages and demand for labor to areas of higher income and opportunity.

The centrality of economic motivation was established in the first systematic study of migration. Ravenstein (1885; see Grigg 1977; Lee 1966) presented his findings as a set of 'laws' of migration (Lewis 1982). Like many nineteenth-century laws in social science, these were empirical generalizations based on Western societies (census data from Britain later augmented with data from several other countries). Not surprisingly, these rules are specific to time and place and reflect the process of industrialization pulling rural Britons into urban work centers. Thus Ravenstein observed mainly short distance migration (although his unit of analysis, the county, varied in size, affecting the actual distances traveled by migrants). Similarly, he found townspeople to be less migratory than rural folk, the direction of migration was primarily from rural to urban areas, most migrants were adults (presumably job-seekers), and migration increased as commerce developed and transport improved. Interestingly, he found females to be more migratory than males.

Grigg's (1977) evaluation of Ravenstein's work points out the role of industrialization in structuring migration in Britain. In so far as similar conditions apply in other regions and times, similar patterns might be expected to occur. Thus young adults are often mobile job hunters and jobs are often in urban centers around the developing world. As Grigg notes, however, in Britain itself, migration to towns increased with 'commerce' but by the 1880s was already declining. The key point is that developing and developed commercial societies often depend on a mobile labor force and the volume of economic movement is sufficient to overwhelm other causes of mobility. To conclude that migration is always economically motivated in the narrow sense of the labor market may miss other important attributes of the process.

Despite these caveats, Ravenstein's work identified several important aspects of human migration and is the foundation for later theory (Lee 1966). Indeed, his view that migration stems from the desire of individuals to 'better themselves in material respects' cited by Ravenstein (1889, quoted in Lee 1966) links his ideas to modern biological definitions (recall Dingle's point that migration is directed toward improving environmental conditions for organisms – Dingle 1996). Subsequent causal investigation of human migration has continued to emphasize economics but has added spatial and social variables as important factors.

Spatial models

Human geographers, not surprisingly, are particularly interested in the spatial aspects of migration (Clark, 1986; Robinson 1996). Spatial interaction forms the core of their discipline and the three basic geographic concepts are distance, direction, and connection (Olsson 1965). These concepts are applied in theories of spatial location (Haggett 1966) as well as models of the diffusion of innovations (Hägerstrand 1967) as well as migration (Olsson 1965).

Of the spatial variables, distance has traditionally received the most attention particularly by geneticists (see Chapter 3). Indeed, in some of the classic models of population genetics, distance is the *only* variable – e.g.,

Malécot's (1955) isolation by distance model relates the decline in genetic similarity to distance alone.

Numerous empirical and theoretical studies relate migration intensity to distance (see Lewis 1982). The intuitive perception that distance acts as a barrier to movement has been amply confirmed and the interest has been in specifying more precisely modifying variables including social and historical effects on the migration–distance relationship.

Consideration of the direction of migration has been combined with distance in the so-called *social gravity* model (Lewis 1982) on the analogy of the force of gravity being proportional to the mass of the attracting body. This relationship can be written as:

$$M_{ij} = K(P_i P_j / d_{ij}^{\,b})$$

where M = migration from place i to j; P = population size of places i and j; d = distance between places i and j; and K and b are constants specific to the situation.

The basic idea is that some places exert special attraction for migrants, which will bias the direction of movement. 'Mass' is represented by the population size of the destination, large towns and cities being more attractive proportionate to their population.

In so far as population size is an adequate operational definition for 'attraction', the gravity model should predict movement. All other things being equal, job opportunities, for instance, should be proportional to the size of a place. However, empirical studies using the gravity approach have produced mixed results (Lewis 1982) suggesting that population sizes may not be sufficient to define attraction. Other size-independent factors may also be involved. Olsson (1965), for example, cites the special pull of warm climates such as Florida or California for migrants hoping to escape winter's misery.

More comprehensive measures of attractiveness of places have been devised to increase the realism of the gravity model. Morrill (1965) for example, employed an index of attraction, A, defined in terms of accessibility of the place to the transport system, level of urban growth, and population density in place of the product of population sizes.

Olsson's (1965) classic study of internal migration in Sweden may serve as an example of the geographic approach to spatial interaction models. He examined variation in migration distances as a function of variation in the characteristics of places along with variation in migrants' demographic or economic status. He went on to consider variation in migration intensity as a function of distance. This approach topic also tied the analysis into a broader locational framework, the hierarchy of central places (Haggett 1966).

In order to test the determinants of migration distances, Olsson (1965) used multiple regression analysis. Characteristics of places included areal size, population size, occupational structure, level of income, and geographic situation; for migrant individuals, typical demographic and sociological variables such as age, sex, marital status, occupation, and income level were used. He found that the distances that migrants traveled was positively related to the level of income of the migrant's origin place, to the level of unemployment in both the origin and destination, and to the population sizes of both places. All of these relationships were significant at the 99 percent level (Olsson 1965:18–20). The characteristics of the migrants themselves such as family income and age were less clearly related to distance.

The second stage of Olsson's analysis treated distance as the independent variable in a gravity model to explain the intensity of migration between places. The salient finding in this work was that the effect of distance on migration intensity varied as a function of the central place hierarchy (Olsson 1965:32). That is, for lower levels of the hierarchy (smaller places), the decline of intensity with distance was steeper than for higher levels. The reasoning behind this result is that central place theory defines a hierarchy of *functions* of places. Large centers are the only places that specialized goods and services may be obtained. These marketing centers require a large hinterland of consumers (range) and a threshold population size in order to provide these functions profitably. Smaller markets serve smaller ranges and provide a narrower range of functions. The spatial pattern of the hierarchy depends on the distance consumers will travel. Daily needs will be satisfied locally, but specialized needs may be provided at lower cost (or only be offered) at more distant larger centers. These advantages outweigh the costs of travel. Relating this back to Olsson's findings, the sharper decay of distance among smaller places may reflect the basic similarity and substitutability between these rather uniform places (Olsson 1965:32). Movement between larger, more economically specialized centers is less constrained by distance and therefore shows a more gradual slope.

Another component of increased attractiveness influencing distance decay in the gravity model is information. Morrill & Pitts (1967) showed that social interaction is strongly influenced by the mutually reinforcing factors of nearness in space and amount of information possessed by individuals about a place. Migration and marriage distances, then, would be influenced by map distance and also by the knowledge of conditions and opportunities available in the potential destination. Nearby places might be expected to be better known than more distant locales accounting for the interdependence of the variables. But, as they (Morrill & Pitts 1967:406) note, 'superior information can overcome great distances'. This basic idea that information held by individuals might determine their behavior was adopted by human biologists in the concept of 'neighborhood knowledge' (Boyce *et al.* 1967) and will be considered in more detail in Chapter 3.

The concern with information fields relates the microlevel processes of behavioral decision-making by individuals or families to the broader patterns of population movement in regions. Focus on the microlevel of analysis highlights the role of non-economic factors such as values, community and social networks, environmental constraints and kinship structure in addition to the usual economic determinants of migration (De Jong & Gardner 1981). Lewis (1982) has reviewed a number of these studies. Theoretical and empirical studies of decision-making with respect to all sorts of behaviors has been presented by C. Gladwin (1989); these models and methods are clearly applicable to the study of migration.

Anthropological genetics: integration of social and biological perspectives

Anthropological genetics, by straddling the disciplinary boundary between social and biological sciences, may relate the socio-economic *causes* of migration to the evolutionary *consequences* of gene flow. Over the course of human history (including prehistory), basic socio-economic features of human populations have varied greatly. The reasons for movement as well as the scale and structure of migration have changed along with cultural patterns. Thus the mass migrations of recent history are the result of the conjunction of particular historical–technological (transport)–economic factors as are the contemporary phenomena of the growth of mega-cities and intra-urban migration. Much current social science research attempts to understand these modern phenomena in terms of locational and economic variables. In addition to these national and international movements, the ethnographic cases to be presented in the next chapter document a wide range of variation in patterns of local migration from extremely sedentary societies with only limited marital exchanges to highly mobile nomads.

The task is to identify the key variables structuring human migration from this great diversity of cultures and histories. As will be seen in the following chapters, patterns of mobility can be correlated with ecological, economic, demographic, and social factors.

2 The anthropology of human migration

Perhaps no other behavior has suffered so much from confusion of definition as migration. What types of movement should be considered migration and how these types relate to one another are complex questions that have resulted in a definition barrier impeding synthesis and generalization across systems. To a great extent this is because students of the phenomenon have tended to focus on a single taxon and its peculiarities, with the result that definitions depended very much on the particular group under study   Dingle 1996:20

Dimensions of diversity in cultural and ecological circumstances

While Dingle is referring to species and higher level taxonomic categories in this passage, the general conclusion also seems applicable to the study of human migration. Beginning with the general 'laws of migration' of Ravenstein (1885) which were generalizations based on the patterns of movement characteristic of one historical period (industrializing countries), models and discussions of human migration and population structure fail to take account of the diversity of ecological and cultural circumstances under which humans have lived and moved. The models considered in Chapter 1, while framed in universal terms, have focused on migration in stratified complex economies with differing employment opportunities. Variables such as 'distance' and 'information' may have applicability across different economic and ecological domains, but most of the empirical examples analyzed in terms of these theories have been 'developed' or 'developing' countries, usually hierarchically organized into central places.

Biological anthropologists and anthropological geneticists have also tended to limit the range of human societies considered in their models. Thus, the populations of South American Indians studied by Neel's multi-disciplinary group were claimed to represent the 'real' human population (Neel 1984); that is, tribal populations similar, demographically and politically, to humans throughout most of our evolution. Cavalli-Sforza's (1986) work on African Pygmies was also represented as 'the earliest way of life'

(Cavalli-Sforza & Cavalli-Sforza 1995). Birdsell (1973) found the 'basic demographic unit' in the Australian dialectical tribe and he (Birdsell 1958; 1968) and Williams (1974), generalizing from Australians and some other hunter–gatherers, identified the 'magical numbers' of 25 persons for bands and 500 for tribes. Such units were explicitly presumed to have characterized human populations throughout prehistory from the Pleistocene to the present (Birdsell 1973:337).

Arguing against the tendency to generalize about all of human evolution from a limited sample of populations is the enormous range of population densities and mobilities shown by various historic and modern human societies. Comparing western desert Australian densities – perhaps averaging one person per 170 km^2 (Cane 1990) – with that of Asian wet-rice farmers with nearly 800 persons per square kilometer (Johnson & Earle 1987) and the mobility of Polynesian ocean-going canoes with foot travel in tropical forests might begin to suggest the dimensions of difference even prior to the modern development of mechanized transport and megacities. To simply assert that 'things are complicated', however, does not advance understanding very far. Ideally, patterns can be discovered in this diversity.

In contrast to the rather limited comparative perspective employed to model human migration and population structure, the search for general patterns in cultural evolution has a long history. By the early 1960s, a sequence of cultural stages representing increasing degrees of socio-cultural integration was in place – bands, tribes, chiefdoms, and states. Derived from the developmental stages of the nineteenth-century cultural evolutionists such as Morgan (1877, reprinted 1963), these categories were given theoretical and empirical substance by Steward (1955), Service (1962), Sahlins (1968) and more recently, by Johnson & Earle (1987).

Johnson & Earle (1987) retain a typology of levels of socio-economic integration similar to the familiar band, tribe, and state classification. Their types include the (1) *family- level group*, (2) *local group*, and (3) *regional polity*.

Family-level groups are found in societies with population densities ranging from less than one person per 10 square miles (*c.* 16 km^2) up to one or two persons per square mile (*c.* 1.6 km^2). These societies include the classic mobile foragers who range as independent families to pursue dispersed resources or who may aggregate into camps of 25–50 persons when food is concentrated. Family groups in more sedentary societies, often with higher population densities, and sometimes relying on some domesticates, may cluster in hamlets. Although less mobile and fluid in composition than the foragers' camps, these hamlets fragment and reform over the short-term as households move in and out. For both camp and hamlet, political organization above the family is ephemeral and context-specific; home ranges are undefended and warfare is rare.

The local group level includes societies with larger populations of many families (up to 5 or 10 times larger than the family level groups) and is found under conditions of higher population density (greater than one person per square mile). These larger groups are based on a common need such as storage of food or defense. Subsistence is usually (but not exclusively) based on domestic crops. Politics may be organized along kinship lines (the acephalous local group) or by a Big Man, a charismatic leader who coordinates internal and external relations.

Regional polities occur when warfare leads to the incorporation of defeated local groups into larger societies held together by an elite. This type of integration may be accomplished by chiefs in societies with economies little different from those managed by Big Man collectivities; however, as economic opportunities for further control arise, and the population becomes denser and more numerous, ruling elites become institutionalized. State organization results from the continuation of this process to include ever larger areas and populations.

This capsule summary of the Johnson and Earle classification is intended to provide a framework for comparing societies' population structures and mobilities in the present work. Thus societies within each level of socio-cultural integration, family, local group, or regional polity, should be similar in numerous respects including population structure. However, as they emphasize, evolutionary change is a *process* involving changes in several variables and societies within each level can be expected to vary quantitatively along a continuum with respect to these key factors.

The first factor is the degree of intensity of the subsistence economy. The primary cause of increasing intensification according to Johnson and Earle is population growth. They see the capacity for population increase to be inherent in animal and human populations, a tendency often checked by environmental conditions but once begun, driving populations to exploit their environments more intensively. Intensification may be achieved through broadening the diet to include less easily obtained foods or increasing work effort and/or investment on cultivation. They see this process leading to changes in the second and third variables – economic and political integration, and social stratification.

Increasing intensification of the subsistence economy generates problems depending on various environmental factors. Johnson & Earle (1987:17) identify four: (1) production risk; (2) resource competition; (3) increasing demand for capital investment in technology; and (4) increased need for trade. Not all of these are inevitable outcomes of intensification but each may give rise to social mechanisms (i.e., increased economic and political integration) to deal with the problem.

Risk at the simplest level can be managed by community food storage or by reciprocal visiting networks among communities to be activated in times of local shortages. However, with increasing population and intensification, political leaders may arise to coordinate and control food allocation. Similarly, as intensification leads to investment and localized valuable resources, greater competition for these valuables may occur. Resource defense is another route to increased political control as alliances for mutual aid acquire leaders. Pooled resources or the need for large scale technology beyond the means of individuals or families along with increasing trade also may lead to political managers and increased political control.

The final variable, social stratification, is based on the degree to which leadership and political control exercised to solve the problems of intensification become transformed into social distinctions among members of the society. With leaders come social classes. As Johnson & Earle (1987:18) phrase it: 'economic intensification creates opportunities for economic control that in turn lead eventually to social stratification'.

The view that population growth is the pressure driving intensification is usually associated with Ester Boserup's work (1965), *The Conditions of Agricultural Growth*, and has been the subject of considerable debate in anthropology (Bronson 1977) and elsewhere. As her title suggests, intensification is normally considered in relation to agricultural practices, the main variable being the frequency with which land is cropped. Boserup specifies a continuum of land use types from virgin land through land cropped with shorter and shorter fallows to continuous cropping. However, there is no intrinsic reason that this continuum can not encompass foragers exploiting 'virgin' land, especially since there is now evidence for foragers practicing various types of land management, such as burning of range to promote plant growth (Williams & Hunn 1982). Such investment practices grade through various degrees of intensifying use of land by swiddening horticulturalists to densely settled, highly intensive agricultural systems such as Asian terraced rice farming.

It is not necessary to agree with Boserup (or Johnson and Earle) that increasing population densities *cause* intensified usage to recognize *correlations* between dense human populations and attributes of technology, economy, and society. The advantage of comparing societies along a gradient of variation in population density and land use intensity rather than by the more traditional technological 'stages' is the explicit recognition of differences *within* categories like 'hunter–gatherer' and 'agriculturalist' as well as similarities between some hunter–gatherers and some agriculturalists in basic features of their population structures and mobilities. To put the argument in a different frame, the difference between *referential* and *conceptual* models in the sense of Tooby & DeVore (1987) might apply to the distinction between ideal types such as 'hunter–gatherers' and a dynamic model of populations with density and intensity as variables.

At the same time, there is no necessity that this dimension of comparison will yield a linear order of increasing or decreasing mobility with increasing density and intensity of land use. Many variables affect migration. Geographic barriers and cultural/ethnic boundaries are classic factors limiting movement. Clearly, the fact that Tristan da Cunha is a tiny island in the midst of a vast sea has a lot to do with the migration into and out of its human population (Roberts 1967). Rivers, mountain ranges, and swamps can all channel or limit movement and affect the isolation of societies. Indeed, circumscription (Carneiro 1970), whether political or geographic, may well lead to higher population densities, increased intensity of land use and even further political barriers to movement. Thus a variety of specific geographic/historical/cultural causes in addition to technology and economy can be expected to play a role in determining the intensity of land usage and mobility. Nonetheless, by employing an explicit framework of comparison, it may be possible to discover patterns in migration across a range of human populations.

The aim of the remainder of this chapter is to survey a series of well-studied human populations to provide a comparative perspective on human migration. The primary dimension of comparison, as discussed above, will be the density and intensity of land utilization by the human population. For each case, the factors affecting population structure and mobility will be examined and special issues illustrated by the case will be detailed.

Table 2.1 presents population data for the several human groups that will be compared and Figure 2.1 shows their locations. Although only a small sample of populations, it has been chosen to represent a wide range of land use intensities from foragers to peasants. An obvious further criterion for inclusion in this sample is that basic population characteristics and migration patterns have been well documented by anthropologists and/or geneticists.

Low population density, extensive land use, family groups

Australia – Yolgnu

Debates concerning the social organization of aboriginal Australia have a long history. This controversy is important to anthropological genetics

Table 2.1. Case study populations: basic demographic data

Group	Year	Tribal/regional population	Local group	Density[a]	Marriage distance[b]
1. Yolgnu	1971–74	2600	20–60	< 0.1 – < 1.0	20[c]
2. !Kung San	1968	2500	20–30	< 1.0	70
3. Aka	1967–76	2000	30–100	< 1.0	53
4. Vaupés	1969	7000	20–60	< 1.0	38.4
5. Yanomama	1970	15,000	25–250	< 1.0	
6. Semai	1968	12,750	20–275	< 5.0	5.3
7. Gainj	1977–83	1500	20–200	c. 25.0	1.3/4.2[d]
8. Basque Baztán Valley	1850–1910	9500	c. 1200	c. 25.0	10.4
9. Oxfordshire	1851	3142	300–400	c. 200.0	9.6–12.8
10. Northern India	1951	1.5×10^6	115[e]	c. 300.0	16.8

[a] Persons per square kilometer.
[b] Mean distance in kilometers between spouses' birthplaces.
[c] 52% of marriages within 20 km.
[d] Father–offspring/mother–offspring birthplace distances.
[e] Families.

References: 1. White (1995); 2. Yellen & Harpending (1972), Harpending & Wandsnider (1982); 3. Cavalli-Sforza (1986); 4. Jackson (1983); 5. Chagnon (1972), Smouse (1982); 6. Fix (1982b); 7. Smouse and Wood (1987); 8. Abelson (1976); 9. Harrison (1995); 10. Gould (1960).

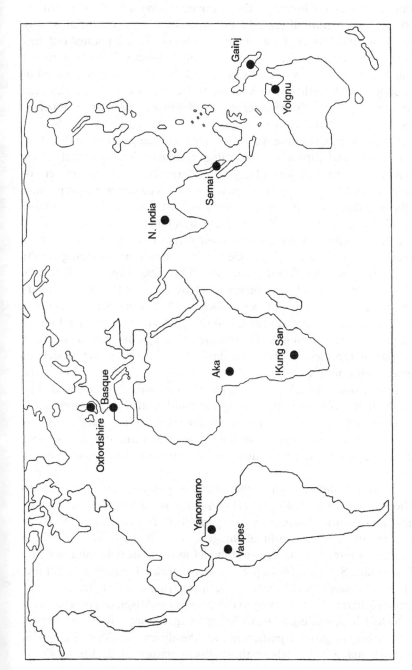

Figure 2.1. Location of case study populations.

because models of Australian population structure have strongly influenced our views of hunter–gatherer organization especially through the
work of Joseph Birdsell (e.g., 1958). Australia has been crucial to any
general model of hunter–gatherers since, as Service (1962) pointed out, this
was the only region of the world where contemporary foragers were not
living in close contact with non-foragers. Only Australia was a 'world of
hunter–gatherers' rather than a mosaic of cultures employing different
technologies potentially influencing and adapting to each other. Particularly if political relations were crucial for structuring groups, as Service
believed, such 'equal footing' among people of similar population densities,
technologies, and capabilities for cultural competition with each other
should provide the best view of any extant foragers of a general pattern of
hunter–gatherer social organization. If Australian groups were not pristine
by the time their social life was described, at least they could not have been
in symbiotic economic relations with surrounding agricultural people for
centuries (Headland & Reid 1989). Their ways of life, then, might reflect
common features of a foraging mode of existence thereby providing insight
into Pleistocene population structure. This was explicitly Birdsell's
(1973:337) view: 'They [Australian tribes] are a model for other cellular-
structured, patrilineal-band types of societies, both those which have persisted elsewhere into present times and those which are presumed to have
been preponderant during the Pleistocene'. Although this view is no longer
in vogue, it represents a powerful implicit backdrop to our models of
human evolution. If we are to understand movement and gene flow in the
Pleistocene, *some* model of human populations must be employed. The
derivation of this model must be explicit and founded on a comparative
understanding of the range of contemporary human patterns. Thus it is
worth examining in some detail both Birdsell's ideas and the broader
anthropological issues and critiques regarding Australian social organization.

A powerful stimulus for constructing a general model of hunter–
gatherers was Radcliffe-Brown's (1930) definitive description of a conti-
nent-wide uniform pattern of social organization. If Australia provided our
best window into the world of foragers, and all Australians shared a
common pattern, then any general model must be heavily influenced by
this structure. Service (1962) especially was concerned that cultural contact
with non-foragers would destroy the underlying political rationale for
hunter–gatherer society leading to the amorphous 'displaced persons' syn-
drome he felt characterized many living forager groups. Radcliffe-Brown,
by describing a general pattern across the diverse ecological zones of
Australia, made more salient the political nature of group structure.

Foragers competing/interacting with other foragers (rather than farmers) provided the foundation of group organization. For Service (1962), the organizing principle was male cooperation in warfare leading to the classic patrilocal band structure. Males upon marriage continued to reside in their natal band and the resulting groups of brothers and other male kin formed an effective fighting unit to oppose other such groups. Here the importance of the political environment becomes obvious; bands would be ineffectual in opposing larger groups of more densely settled farming populations.

The basic pattern that Radcliffe-Brown presented for Australia was divided into three levels – the family, horde, and tribe. The family was merely the familiar nuclear unit, a man, his wife, and children. But as Radcliffe-Brown (1930:34) noted, 'it is not easy to give a precise and accurate account of the local organization of Australia'. Nonetheless, he provided an unequivocal definition of the horde as territorial, patrilineal, exogamous, and politically autonomous. These small groups of related males with their spouses and offspring were the owners of the land and resources and the day-to-day economic units of Australian society according to Radcliffe-Brown. The tribe was simply a collection of hordes speaking dialects of one language. The name of a tribe and the language they spoke were normally identical. Essentially, then, the tribe was not a political unit but a linguistic group that possessed a shared culture as well.

The most controversial of these three categories has been the horde. It was the horde which formed the basis for the patrilineal/patrilocal band models of Steward (1955) and Service (1962) and has been the focus of contention in Australian ethnography ever since Radcliffe-Brown's pronouncement. The difficulty with discovering local group organization lies partially in the complexity of the relationships of Australians to their country (see Hamilton 1982, for a particularly complicated discussion) and partly (as Radcliffe-Brown pointed out) due to the destruction of the aboriginal system of land tenure and social arrangements following European colonization and expropriation. Where Radcliffe-Brown saw a uniform pattern, many later ethnographers found diversity (see Hiatt 1996, for a summary).

A recent description from Arnhem Land in the northeast of Australia provides a contrast with the Radcliffe-Brown model (White 1995). White states that the Yolngu (also called 'Murngin') 'are organized into named land-owning patrilineal descent groups (clans) and territorially based dialect units' (White 1995:254). As for many other areas of Australia, land 'ownership' involves stewardship of sacred sites which are linked to other such sites in the territories of other clans along the tracks of the mythical ancestors. Both clans and dialect groups are ideally exogamous. The actual land using groups, however, include family groups from one or more clans with foraging rights by virtue of kinship (consanguineal and affinal) over the lands of several other clans. These groups vary in size with season and habitat and are flexible in relation to the availability of resources.

White goes on to point out that this flexibility is particularly crucial in the more arid regions of Australia. Thus for the Pintubi living in the desert with population densities lower than one person per 200 km^2, long-distance social links through marriage and religious affiliation allow access to different resource areas in time of need (White 1989). Similarly, Tonkinson (1978:161) reports for a western desert population that 'kinship, ritual, marriage alliances, shared ideology and a host of other cultural elements stress broader linkages which allow for the retention of access rights in the territories of other groups'.

White (1995:297) concludes that 'Australian Aboriginal population structure varied greatly throughout the continent', counter to Radcliffe-Brown's claimed universality for the horde. With coastal population densities considerably higher than those in the desert (from one person per 3 km^2 for coastal Yolngu decreasing to one per 16 km^2 for inland Yolngu to the extremely sparse interior desert population of the Pintubi at one per 200 km^2), differences in foraging patterns, marriage distances, and stability of local groups should be expected.

Although local group structure may affect marriage distances and patterns of mating, hordes or bands should not constitute actual breeding populations since ideally they were exogamous entities. The deme by definition is an endogamous grouping. Thus the tribe was the 'basic demographic unit' for Birdsell (1973). He argued that the linguistically-defined tribe was kept in stable equilibrium numerically at approximately 500 individuals and was relatively endogamous (outmarriage rate of about 15% per generation). The basis for tribal cohesion was not political integration but rather the social interactions taking place among members of the local groups comprising the tribe. Above a certain size (500 persons), the density of face-to-face contact is insufficient to maintain a common dialect. Regional linguistic divergence creates new smaller tribes and, since Birdsell has demonstrated to his satisfaction that linguistic tribal boundaries inhibit marriages beyond merely isolation by distance (Birdsell 1958), new breeding populations.

The Yolgnu seem to fit Birdsell's tribal breeding population parameters. White (1995) derives an effective population size of 550 for the tribe as a whole (based on a census size of 2600 persons of all ages) with a very high level of endogamy (96%). However, as he also points out, for the arid regions of Australia the very fluid local groups result in permeable boundaries between tribes 'leading to the establishment of chains of dialects and genetic clines' (White 1995:258). It may well be that the degree of 'tribal' cohesion and boundedness varies with population density among Australian groups and within the technologically defined category, hunter–gatherer. The coastal Arnhemlanders with higher population densities are more 'tightly packed', having more rigid social boundaries and smaller breeding populations (White 1989).

White (1989) points out a further factor structuring Aboriginal populations, river basin drainages. Yolgnu subgroups are localized within stream and river drainage basins where food resources are concentrated. Moreover, river courses are the main routes of travel in dry Australia and thus groups found along a drainage will be linked. This contact leads to increased marriage within the basin. It might be noted that the linear, reticulate pattern of rivers might reduce the dimensionality of migration and gene flow.

South Africa – !Kung San

The San of the Kalahari Desert are perhaps the best-known contemporary foraging people as a result of the long-term studies of Marshall (1976) and the Harvard Project (Lee & DeVore 1976). As with the Australian material, analogies and models derived from the San have influenced inference about biological process. Indeed, Glynn Isaac's (1978) 'carrying and sharing' model seemed to be a direct extrapolation of !Kung social patterns to the ancestral hominids. Early hominids could be thought of as simply '!Kung in the Pleistocene'. Characteristic attributes of modern !Kung including carrying food back to a home-base to be shared were central features of Isaac's model accounting for the evolution of early human biology, culture, and behavior.

!Kung local group patterning has also been suggested as a model for hunter–gatherer population structure with implications for archaeologists and biological anthropologists (Yellen & Harpending 1972). In particular, Yellen and Harpending take issue with the band as a useful concept for the on-the-ground population units of foragers. They argue for many hunter–gatherers that resources are very heterogeneous and scattered in distribution and that social organization is extremely flexible in order to allow efficient use of these varied plant and animal foods. Under these circumstances, tightly closed groupings of unilineal kin (i.e., the patrilineal band) would limit the ability of individuals to respond to this environmental heterogeneity. In their view, a band of !Kung is 'no more than a temporally unstable aggregation of families with links of kinship or friendship functioning primarily to smooth the daily variations in exploitative success of its members by food sharing' (Yellen & Harpending 1972:247). Thus food sharing (and information sharing about the location of food items) provides an advantage for group membership. The scattered nature of most food resources, their usually rapid local exhaustion, and the lack of political mechanisms of dispute resolution limits group size and causes continual fragmentation of groups. Over the long-term, then, families split and come together in keeping with seasonal and annual local variation in resources.

This view of local group 'organization' has been called 'anucleate' by Yellen and Harpending. They describe !Kung social patterns as a network of ties of kinship, friendship, and trading, extending widely across space such that individuals have rights to resources in many different areas. As resources vary, families or individuals may move to 'visit' with kin, friends, etc., in areas of relative abundance. Many aspects of !Kung sociology can be related to this widespread extension of amity e.g., the 'name-relationship' in which kin terms are applied to individuals with the same name irrespective of their actual affinal or consanguineal status (Marshall 1976), and the *hxaro* relationship, a kind of social storage of debt allowing partners visiting privileges in times of need (Wiessner 1982).

Harpending & Jenkins (1974) found that !Kung endogamy, mating distances, and genetic microdifferentiation were all consistent with the anucleate nature of local groups. Dividing !Kung land into nine areas of approximately 200–300 persons (a population size similar to many small scale horticultural societies), they found that the proportion of parents in an area that were also born there (a measure of endogamy) to be 0.57, a low figure compared to more nucleated societies such as the Maya (0.97) and Oxfordshire villagers (0.83). Likewise, the average distances between birthplaces of spouses was 70 km for the !Kung, greater than but comparable to the 53 km distance of the Aka Pygmies of the Central African Republic (Cavalli-Sforza & Hewlett 1982) and much greater than for more settled agricultural people. Thus social nucleation provides a dimension of comparison along which populations may be compared. Hunter–gatherers such as the !Kung with relatively anucleate settlement patterns should be less locally differentiated; even the short-term breeding population will have much larger areal extent than more nucleated populations.

What is not clear from the analysis of the !Kung is whether there are barriers to mating comparable to those suggested by Birdsell (1973) for Australian tribes. Potentially the !Kung are less likely to mate with other San groups than they are with other !Kung. On the other hand, if mating networks extend seamlessly across dialect boundaries, then the 'deme' extends to the entire region or beyond. Anthropological geneticists are so accustomed to neat socially-defined populations in more densely settled agrarian cultures that the concept of a continuum of mobile individuals and families extending over very large spaces may be hard to accommodate. It may be that for such anucleate societies, Wright's (1943) isolation by distance model is more appropriate than those presuming subdivision into local breeding populations. Certainly such ideas ought to be considered as we attempt to understand human genetic history over the long term.

Central Africa – Aka

The numerous small-statured peoples of Central Africa variously called Twa, Aka, Binga, Efe, and Mbuti have represented the ideal type of tropical forest foragers to generations of anthropology students. Particularly through the writings of Colin Turnbull (e.g., 1961), the Pygmies epitomized hunter–gatherers who, although engaging in mutually advantageous interaction with settled farmers, were essentially free and self-sufficient in the forest. This idea led the geneticist Cavalli-Sforza (1986:378) to study their population 'in the hope that it might give us some idea of conditions in the Paleolithic'. In fact, data from the Aka influenced Cavalli-Sforza's (Rendine *et al.* 1986) models of European Mesolithic foraging population structure (see Fix 1996).

Since the 1980s, various revisions of this view of Pygmy history and autonomy have appeared (Bahuchet & Guillaume 1982) including an argument that foraging in the tropical forest without benefit of trade with farmers for carbohydrates would be impossible (Bailey *et al.* 1989). But leaving aside the question of whether or not the Aka represent the 'earliest way of life' (Cavalli-Sforza & Cavalli-Sforza 1995), they can provide comparative information on mobility and migration in a low density population.

Cavalli-Sforza (1986) discusses the hierarchy of population clusters among Central African Pygmy groups. First, as he points out, the largest unit, the 'tribe' or dialect group is an 'unsettled problem' in the case of the Pygmies. In contrast to Australia, language groupings (the dialectical tribe) can not be easily identified because Pygmy groups speak variants of the languages of the farmers with whom they are associated. Thus various Pygmy groups speak languages from several different Bantu subdivisions while others speak Sudanic languages. The adoption of the languages of their farming neighbors indicates a long and strong relationship between

Pygmies and more settled people. This symbiosis, in which Pygmies supply game and occasional field work in exchange for carbohydrates, may also affect the structuring of regional bands as groups space themselves not only with respect to natural resources but also to the 'social' resources represented by settled villages. To what extent population density is affected by farmer–forager interactions is an open question since no known Pygmy group lives isolated from farmers. The rather low population densities recorded by Cavalli-Sforza (1986:369) of 0.1 to 0.5 persons per square kilometer (the same order of magnitude as the !Kung San – see Table 2.1), suggest that the 'extra' resources provided by the agriculturalists do not have a major impact on the Pygmies.

Cavalli-Sforza (1986:361) estimates the total population of African Pygmies to be 150,000 to 200,000 occupying forested regions of Central Africa. A variety of names are applied to groups in different areas of this wide distribution. The Akas of the Central African Republic studied by Cavalli-Sforza's group number more than 3000; similar groups in the Ituri Forest of the Republic of the Congo are even more numerous (perhaps 10,000 to 15, 000). As noted above, subdivisions of these very large aggregations into 'tribal' units is problematic. Cavalli-Sforza, however, defines a hierarchy of groups functionally in terms of (1) coresidence, (2) hunting, (3) marriage, (4) social interaction, and (5) communication.

The coresident group averages 30 persons occupying a camp which frequently (every two months on average) changes location. The composition of this group may also change through time. Turnbull (1968; 1986) has argued persuasively for the *political* importance of flux in such local groups in the Ituri. Camps fission and fuse in keeping with seasonal changes in hunting and honey-gathering. But these ecological factors are merely triggers that allow the pent up tensions developed between families and individuals to be defused by fission of the camp. After a period of separation, groups reform at the beginning of the new season sometimes with the same members or often with others. This fluctuating local camp composition is similar to the already discussed case of the !Kung. Clearly camps are not in any sense breeding populations.

Cavalli-Sforza additionally defines a 'hunting band' separate from the camping unit. He notes that members of a camp usually forage together; however, hunting groups vary in their requirements for personnel. Thus cooperative net hunting requires at least seven nets (i.e., seven families since each family owns one net) and larger groups up to 20 nets are optimal. In order to achieve these larger parties, families from several camps must come together. Even more than camps, these 'task groups' (Helm 1968) are ephemeral and are inappropriate units among which to measure migration. The low population density and the high mobility of individuals making up these groups ensure that a large geographic range is familiar territory to the Aka.

Marriage serves also to extend the spatial spread of Pygmies. Turnbull (1986:113) says that marrying 'far' is the major consideration for the Mbuti. The primary reason for wide marriage ties according to Mbuti is that having affines in distant bands provides refuge when fission follows dispute, although economic advantages are also cited. This prescription results in most marriages occurring between individuals in bands separated by two or three territories up to 50 miles (c. 80 km) away. Cavalli-Sforza (1986) reports a mean marital distance of 53 km for the Aka.

The lability of anucleate groups and the great mobility of individual Aka, as for the !Kung, make determining the boundaries of local breeding populations problematic. Cavalli-Sforza (1986:384) attempts to overcome this obstacle by defining a statistic, the *mating number*, which is 'the number of individuals from among whom a spouse is chosen'. This measure is similar to Dyke's (1971) concept of a 'potential mate pool', which depends on specifying the availability of opposite-sexed individuals of marriageable age and kin status for an average member of the population. Mating number (N_m) as defined by Cavalli-Sforza, however, is simply

$$N_m = \Sigma p_i n_i$$

where p_i is the probability of marrying an individual from the *i*th population unit, and n_i is the number of residents of that unit.

By dividing the Aka of the Central African Republic into nine areas (arbitrarily defined simply on geographic location), and by using parent–offspring birthplace data to approximate marriage probabilities, Cavalli-Sforza found mating numbers to vary between 564 and 1616 (the average for all nine was 870).

This formulation is appropriate for a continuous population distribution. In contrast, if the nine geographic areas are treated as discontinuous subdivisions, average population size is 379 and migration among them is on the order of 38 percent; more inclusive population units with only three subdivisions yield an average N of 1137 and a correspondingly lower estimate of m of 22 percent. The classic population genetics parameter for assessing the role of genetic drift, $N_e m$, the product of the effective (breeding) population size and the migration rate, is 48 in the first case and 84 in the second. Since the average population size based on the mating number lies between these two population estimates (890 compared with 379 and 1137), Cavalli-Sforza simply interpolated to calculate an intermediate value for $N_e m$ of 71 for this case.

Aka migration will be further discussed in the next chapter in the context of models of neighborhood knowledge and exploration range.

Low to moderate density, extensive cultivators, local kin groups

Lowland Amazonia – Vaupés

The complexity of locating boundaries of social units in low density human populations can be illustrated by the regional system of marriage in the Vaupés area of Amazonian Columbia described by Jackson (1983). Particularly interesting given the usual practice of defining population units on linguistic criteria is the striking linguistic differentiation in the region combined with a 'remarkable degree of cultural homogeneity' (Jackson 1983:5). Indeed, the subtitle of Jackson's book, *Linguistic group exogamy*, explains this situation. Jackson (1983:6) maintains that regional systems of interaction occur among widely dispersed local groups within 'virtually all low-density populations around the world, especially hunter–gatherers'. She explicitly points to population density as the crucial variable along with exploitation of extensive ranges that favor 'local group interdependence, fluidity in territorial boundaries, and fluctuations in local group membership' for both classic hunter–gatherers and the people of the Vaupés.

The Tukanoan speakers of the Vaupés see their region as unbounded; people are not hierarchically divided into population units. Groups satisfying the criteria often recognized for tribal definition, that is, with bounded territories, shared language and culture, and endogamy, are not found in the Vaupés region (and perhaps not among other dispersed populations such as the San). Indeed, the language groups used by Jackson as analytical units are exogamous and are not differentiated politically or culturally.

The social system in the Vaupés comprises speakers of more than 16 languages each associated with a descent group. These groups are horticulturalists occupying settlements along rivers in the tropical lowland environment of the Columbian Amazon. Manioc is the most important crop although, like most swiddeners, numerous other plants including plantains, bananas, sweet potatoes, melons, and yams are interplanted. Hunting, fishing, and gathering are also important to subsistence. As suggested already, population density is very low, approximately 0.2 persons per square kilometer. Total population for the region in the 1970s was estimated to be between 7000 and 12,500, although Jackson points out that a long history of disease and exploitation may have reduced the population size and density from former times. Settlements may be a single longhouse with 20 to 40 inhabitants or nucleated villages of up to 60 persons (these latter in response to pressure by missionaries). Post-marital residence is ideally patrilocal. Settlement populations are comprised, then, of an agnatically related descent group along with their in-married wives. These women's ties to their own natal local descent groups remain strong, creating links between different settlements.

Settlements, by virtue of being localized descent groups, are exogamous. This contrasts with some other small scale horticultural societies such as the Yanomamo (see below) whose villages normally contain two or more kin groups that intermarry allowing endogamy (Chagnon 1972). Another obvious contrast with the Yanomamo that has important ramifications is the lack of warfare among the Tukanoans (at least in recent times). However autonomous Vaupés settlements may be, exogamy produces a fundamental dependence on other settlements for wives. These ramifying connections mean that all Tukanoans are in some sense kin throughout the region.

Visiting between settlements is common however long and difficult the canoe trip may be (Jackson 1976). Reasons for visiting include trade, ceremonies, and courtship, and visits may last for some time. As for the Yanomamo and Semai (see below), quarrels can lead to settlement fission with dissident groups going to 'visit' kin in another settlement and ultimately fusing with the group. Jackson (1976) notes that ordinary visiting results in an impressive knowledge of the geography of the region and also influences marriage channels. Settlements thus comprise nodes in a network of trade, marriage exchange, feasting, and (in the past) military alliance extending over large distances.

The 16 language groups are named patrilineal descent groups as well – each is composed of from six to more than 30 sibs. Ideally, the sib is the local descent group resident in a single longhouse making up a settlement although this pattern is not always realized (Jackson 1976:78). Language groups do not occupy a single territory. Although they are generally confined to specific areas, some member settlements are interspersed among others. As with any small-scale system, population size of the language groups varies considerably (from 50 to 2000).

The dispersed population, low population density, and pattern of language group exogamy leads to high mean marriage distances. Jackson (1976:85) graphs distances between birthplaces of spouses ($n = 635$ marriages) based on the linear distance as the crow flies on her map of the region. Using this rough measure, she found the mean marital distance was 24 miles (with a standard deviation of 22 miles; or *c.* 38 km and 35 km respectively). The graph shows a decreasing frequency of marriage with distance but the decline is not exponential and a substantial fraction of marriages occur between spouses whose birthplaces are up to 75 miles (*c.* 121 km) apart. The mean distance of 24 miles (*c.* 38 km) is less than the !Kung but is still substantial.

Lowland Amazonia – Yanomamo

The long-term multidisciplinary study of South American Indians, particularly the Yanomamo but also including the Xavante, Makiratare, and others, by James Neel and his many colleagues (reviewed in Neel 1978) provided one of the most complete descriptions of any human population. This project was explicitly designed to document the population structure of tribal humans, the 'real' human population (Neel 1984) which characterized most of our evolutionary history. Neel argued that neither the large amorphous populations of contemporary nation-states nor the apparently tightly closed peasant populations of Europe were appropriate models for the conditions structuring and shaping human genetic variation.

In addition to the extensive biological data collected by Neel's team, the population structure and mating patterns of the Yanomamo have been documented in great detail based on many years of field research by the social anthropologist, Napoleon Chagnon (e.g., Chagnon 1968). Contrary to the common anthropological practice of studying single communities, Chagnon was initially interested in tribal politics and therefore studied a region containing several villages. Later, working with Neel, this regional perspective was extended to numerous villages over large areas of the tribe. Again, the collaboration with Neel encouraged the extension of Chagnon's already extensive collection of genealogies and village histories allowing a very detailed look at patterns of migration between local groups over a wide area.

A further attribute of the Yanomamo emphasized particularly by Chagnon (1972) was the retention of their freedom from domination by outside political forces, a sovereignty lacking for most tribal people in modern times. Just as the pre-colonial absence of farmers in Australia made it a crucial case for a general theory of hunter–gatherer politics and population structure, it was argued that the Yanomano provided one of the few remaining instances of tribal politics uncontaminated by external domination. For Chagnon, warfare or the threat of warfare among autonomous villages, structures nearly all tribal institutions. Therefore the Yanomamo offered one of the few opportunities to understand the tribal pattern and thereby the social and population structures typical of pre-state formations.

Compared to the hunter–gatherer Australians and San, the swiddening Yanomamo have higher population densities and are more nucleated. Chagnon (1972) estimated their total population size at between 10,000 and 15,000 distributed in 125 widely scattered villages in southern Venezuela and northern Brazil. The overall population density is less than one person per square kilometer (Chagnon 1968) although as Neel (1978) notes, this is a somewhat uncertain measure given the patchy distribution of the Yanomamo. Even in the central more densely occupied region of the tribe, however, Neel estimates only one person per three or four square miles. This figure overlaps with the coastal groups of Australia and is probably considerably lower than the densely settled acorn-gatherers of California or the salmon-fishers of the Pacific Northwest Coast.

Yanomamo villages are generally small, averaging between 90 inhabitants in the center of the tribal distribution to about 50 in the periphery. For the center, the range is between 40 and 250 inhabitants. The lower limit of village size seems determined by problems of defense – a group fewer than 40 persons is extremely vulnerable to raids and would be likely to amalgamate with another group; larger groups are politically unstable and fission occurs. At the edges of the tribe, warfare is less intense and village size may fall to as few as 25–30 persons and rarely reaches 100. These groups are more widely spaced and avoid threats by migrating away from neighbors, an option not available to central groups that must depend on size and a system of inter-village alliances.

Yanomamo horticulture is based on vegetatively propagated crops, primarily plantains but also including manioc as grown in the Vaupés discussed in the previous section. These tree and root crops do not require intensive cultivation and the Yanomamo are not tied tightly to particular plots of land. They clear new gardens every year and return to the same areas only after long fallows. Villages are moved every three to five years usually 5 to 10 miles (*c.* 8–16 km) away (Neel 1978).

The relatively low population density of the Yanomamo insures an ample supply of new land for fields. Their extensive cropping system reduces their commitment to particular plots of land and reduces the cost of movement. Mobility, then, is an easy solution to resolve disputes just as was previously noted for the Pygmies.

Political authority is weak among the Yanomamo. Headmen exist but kinship and consensus are the primary factors maintaining peace within villages. When these mechanisms fail to bring parties to agreement, the only recourse is for dissident groups to leave the village. Since disputes tend to increase as population size increases, and population growth is continually occurring (Neel & Weiss, 1975 estimated the annual rate of growth

[*r*] to be between 0.5 and 1%), the conditions for village fission are endemic.

Continual fissioning of village populations results in both the periodic founding of new village populations as well as the fusion of some components of the splitting groups with already established villages, a pattern Neel and Salzano (1967) called *fission–fusion*. They also pointed out that the social organization of the Yanomamo is based on patrilineal kinship. Disputes tend to polarize between patrilineages; when splits occur, fission groups are thus likely to be kin. Newly founded villages are often established by such kin groups, referred to as *lineal effect* (Neel 1967). This pattern of fission–fusion and lineal effect was shown to produce a striking degree of local genetic variation (Neel & Salzano 1967).

Migration among the Yanomamo, then, is strongly conditioned by both internal and external political factors. Internal village politics leads to disputes and periodic fission while external politics constrains the possibility of migrating and the destinations of migrants.

Along with group migration resulting from village fission, a shifting pattern of marriage alliances between villages produces a 'trickle' of individual marital migrants. Augmenting these peaceful exchanges, some inter-village movement is involuntary as individuals are captured in warfare and incorporated as mates (Chagnon 1972).

Continual fission establishing new villages may generate a branching tree diagram of descent reminiscent of the phylogenetic trees of evolutionary classification. If splitting into new populations was the only process operating in Yanomamo population history, a tree might be an appropriate model. However, the fusion of groups and the background trickle of marital migration and abduction serves to reconnect separate branches producing an anastomosing pattern. Moore (1994) has used the metaphor of river channels shifting and recombining to characterize the process of ethnogenesis and this notion also seems appropriate for Yanomamo fission–fusion.

Among the Yanomamo, villages provide nuclei and serve as the units among which migration occurs. For the hunter–gatherer Australians and San, no such focal group existed. Local 'bands' were continually rearranging, on a scale of days or weeks, as families and individuals reassessed their foraging needs. Yanomamo local groups have more temporal stability but are nonetheless ephemeral on an evolutionary time scale. Endogamy, measured by Chagnon (1972: 272) as adherence to the prescriptive marriage rule, ranged from 71 percent in a large village to only 41 percent in a smaller (*n* = 51) village. Clearly, these are not long-term breeding isolates. Neel (1978) has suggested that over the relatively short-term the 'breeding unit' may be a cluster of 4 to 10 related villages often descended from a common ancestral village through successive fissions and united by continued fusions and intermarriage (see also Smouse 1982).

Malaysia – Semai Senoi

The Semai Senoi are a tropical swidden-farming people of the central Malaysian Peninsula (Dentan 1968). As with the Yanomamo, Semai live in small villages of 20 to 275 persons spread over a relatively large area. In 1965 their population numbered some 12,750 distributed among more than 200 settlements with a population density of roughly two or three persons per square kilometer (Fix 1982b). In contrast to the Yanomamo, however, Semai do not engage in warfare; in fact, they have become well known in the anthropological literature as an example of a nonviolent culture (Dentan 1968). Another contrast with the Yanomamo is the lack of patrilineal organization among the Semai. Whereas Yanomamo villages tend to be comprised of two or more patrilineages and cross-cousin marriage is prescribed, Semai settlements are made up of a series of hamlets of varying size each inhabited by two to five bilaterally related families. Kinship ties ramify through the settlement population and beyond to neighboring settlements. Lacking unilineal kin groups, the primary Semai marriage restriction is with any member of an individual's bilateral kindred (Dentan 1968). At the same time, there is some propensity to marry within kin groups with which affinal relations have already been established (Benjamin 1986) and, in so far as potential spouses are not closely consanguineally related, to marry within the settlement. Both of these preferences may have to do with a mistrust of 'strangers' – in-laws and co-villagers are familiar and therefore less to be feared. Although Semai would prefer to live with kin and therefore settlements ideally are kin groups, in practice, the lability of Semai local groups means that nearly all settlements include unrelated families from which potential spouses may be chosen (Fix 1982b).

Thus the Semai share a number of features with the Yanomamo and differ in others. Both depend on swidden gardens for subsistence and both are relatively sparsely distributed populations lacking strong political control to resolve disputes. However, since the Semai lack warfare and unilineal kinship, the importance ascribed to these factors in determining Yanomamo patterns of marriage and movement can not apply to the Semai case. Nonetheless, the pattern of settlement fission into migrant groups of relatives that either fuse with an established village or found a new one (lineal effect), which is characteristic of South American Indian groups such as the Yanomamo and Xavante, is also found among the Semai. This limited comparison suggests that factors other than feud and lineal descent groups cause this pattern (Fix 1975).

Fission of Semai settlements can occur for a number of causes. Before considering the actual causes of fission, however, it should be pointed out that movement is relatively inexpensive for Semai. Generally their swidden cycle involves only one year of cropping (dry rice and numerous other interplanted crops) though manoic may continue to be harvested for another year. Fallows are variable but usually long and may be as much as 30 years. Once the rice is harvested, there is little commitment to a particular piece of land. New swiddens will be cut for the next year, and this provides a convenient time for people to migrate to another area (Dentan 1968). Furthermore, land is not considered to be a scarce commodity by the Semai, and particular fields or plots are not directly inherited from generation to generation. Many social avenues grant access to the use of land including any bilateral kin link, though in practice people generally live in areas with numerous relatives. The point is that they are not limited in their ability to move to new areas particularly if they have kin already resident in the new locale. Since previous moves have widely distributed ties of kinship through a region, most people have some link to another area if they wish to take advantage of it. The relatively low population density and availability of potential garden plots in many areas makes migration of groups an easy option for whatever reason.

One precipitating factor leading to a fission may be an unresolvable dispute within a settlement. The remarkable nonviolence of the Semai does not preclude arguments and conflict, only that these disputes may not be resolved by fighting. Most can be settled by prolonged discussion by all the members of the settlement that wish to participate. These community discussions are mediated by Semai elders and can sometimes go on for several nights (Robarchek 1979). However, since the elders have no real authority but can only work for consensus, when the differences are irreconcilable, the only resort is for one of the factions to leave the settlement. Thus, settlements may fission among the Semai for similar reasons to those already described for the Yanomamo.

Another factor often cited by Semai for settlement fragmentation is an abnormal number of deaths in a settlement (Fix 1982b). A severe disease outbreak striking a settlement produces a crisis of fear – Semai seem to respond to crisis by fleeing. They burn their houses and abandon the site where death occurs. With many deaths, the population may simply split up and leave the area to set up new settlements or fuse with neighboring ones.

The splinter groups which form from settlement fission are, as for the Yanomamo, kin-based. In the Yanomamo case, the structure of these kin groups is patrilineally biased. The Semai lack patrilineages but the group is no less kin-structured for being bilateral. The critical effect of kin-structuring on the genetic composition of the migrating group is to reduce the effective number of independent genomes since biologically related individuals share alleles identical by descent from a common ancestor (Fix 1978; Neel 1967). This sharing holds irrespective of the kinship reckoning of the culture.

This kin-structured group migration in both the Yanomamo and Semai contrasts markedly with the usual view of random migrants trickling among demes (see Chapter 2). Neel (1967) and Smouse *et al.* (1981) pointed out how lineal effect (kin-structured groups) could lead to marked genetic microdifferentiation as new villages were founded by non-representative samples of the donor population gene pool. Using the same logic and based on the Semai material, I (Fix 1978) proposed that groups of kin-structured migrants that fused with established villages could also increase genetic microdifferentiation.

Semai settlements, like Yanomamo villages, are not strictly demes since only the largest can be endogamous even in the short term. Table 2.2 gives population sizes and birthplaces for spouse pairs (Fix 1982b). These values vary greatly among settlements and reflect population size but more importantly the history of fission and fusion in the region. Small, recently founded settlements such as KE have few parents born in the new locale and thus both for parent–offspring and spouse pair concordance is low. RU is a larger settlement but was recently augmented by a large migrant group from outside the study area (although only about 10 miles, 16 km, distant – see Fix, 1975 for a map and Fix & Lie-Injo, 1975 for further details). SA is the largest settlement in the area and has a long history of interaction with KE and CH and for these settlements, parent–offspring birthplaces are highly localized within the study area (92% in the case of SA).

If settlements are not very endogamous, the larger area defined by all seven does seem to include a higher percentage of spouse pairs' birthplaces. The weighted mean value of spousal endogamy for the study area is 0.81, a figure comparable to Birdsell's (1973) basic demographic unit (85% endogamous) and to Adams & Kasakoff's (1976) values for a variety of endogamous populations. However, it must be realized that the study area represented by these seven settlements is not a 'natural' unit in Semai social organization. Similar to the Yanomamo, Semai villages are politically autonomous. Groupings of settlements occurring along the same river valley, a natural avenue of travel, may have closer relationships but BU

Table 2.2. *Semai: birthplaces of married pairs*

Settlement	N_r	N_s	Settlement endogamous	Region endogamous	Region exogamous
SA	272	117	0.453	0.888	0.112
KE	69	19	0.210	0.946	0.054
RU	107	37	0.054	0.648	0.352
KL	54	34	0.147	0.824	0.176
BU	107	33	0.182	0.847	0.153
CH	117	79	0.304	0.721	0.279

N_r lists the 1969 population of residents in each of the settlements. N_s is the number of spouse pairs where one spouse (arbitrarily husband or wife) was born in the settlement. Settlement endogamy occurs when both spouses were born in the same settlement; region endogamy refers to both spouses being born within the region defined by these six settlements.

settlement has historically had a close tie to settlements on the opposite flank of the main mountain range dividing the Peninsula and Semai country. CH and SA on the other hand are not in the same drainage yet have very close links.

On a larger scale, Dentan (1971) has identified regional clusters and Diffloth (1968) has recognized several distinct dialects of the Semai language, which might form endogamous units. Since the Semai as a whole are defined only in very loose terms as the 'aggregate of people who speak dialects of the Semai language' (Dentan 1968), it is unclear how effective the 'tribal' boundary is to mating. The Semai are similar to Birdsell's linguistically defined tribes in having no political reality but the Australian groups were much smaller units ($n = 500$) than the entire Semai population ($n = 12,500$). Since in historic times, much of the periphery of Semai territory has been occupied by Malays (later Chinese immigrants), data on inter-'tribal' marriage are not available.

Interestingly, the size of the study area population ($n = 776$) within which over 80 per cent of mates are obtained is close to the 'basic demographic unit' ($n = 500$) of Birdsell (1973). Over the short microevolutionary term of one to three generations, the village clusters of the Yanomamo and Semai might be appropriate analytical units, the populations among which migrants move.

New Guinea – Gainj–Kalam

The Gainj and Kalam are forest swiddeners living on the fringe of Papua New Guinea's (PNG) central highlands. They are similar to the Yanomamo and Semai in depending on gardens supplemented by hunting and gathering. The Papuans, however, also keep pigs and chickens, domesticated animals not available to the Yanomamo and not very important to the economy of the Semai. Staple crops are sweet potatoes, taro, and yams, vegetatively reproduced roots rather than grains. Population density in the PNG groups is considerably higher than that for the Yanomamo and Semai, that is, on the order of 25 persons per square kilometer (Smouse 1982) versus fewer than one and fewer than five for the Yanomamo and Semai respectively. Local group (called 'parishes' in PNG) sizes are quite comparable to the other swiddening groups, ranging between 20 and 200 inhabitants. Settlements are less nucleated than for the Yanomamo, comprising one or more hamlets of a few houses each scattered in the parish range. This patchy distribution of hamlets is characteristic of the highlands of New Guinea. Large Semai settlements also often are subdivided into hamlets separated by up to a 20 minute walk.

The parishes are not highly endogamous. Individuals attempt to spread marriage ties widely. The goal is to have affinal connections with as many other parishes as possible (Smouse 1982). Since post-marital residence is preferentially patrilocal, marital movement is mainly by women (Wood et al. 1985). Thus some 84 percent of children are born in the natal parish of their father's compared with 33 percent for the mother. This is a much higher rate of exogamy than for the Yanomamo but is comparable to that of the smaller Semai settlements. Perhaps because of the higher population density, mean marital distances are much lower in New Guinea than for the Yanomamo or Semai. Again expressed as parent–offspring rather than spousal differences, the mean distance between birthplaces of Gainj–Kalam fathers and offspring is only 1.3 kilometers and for mother–offspring 4.2 kilometers. Distance has a strong negative effect on movement partly due to the difficulty of traveling on foot in rugged country. Some idea of effect of terrain on travel is provided by the differences between travel distance (measured by a pedometer along normal trails) and map distance – the most distant parishes are some 40 kilometers apart by trail although maximum map distance is only about 13–14 km (Wood et al. 1985).

Sex-specific dispersal has been an enduring feature of classic band society models; indeed, the defining characteristic of hunter–gatherer bands was patrilocality (Service 1962). However, in the present sample of populations, strict patrilocality is typical of neither the San nor the Semai and, as Turnbull (1986:110) says, the high degree of mobility for the Mbuti makes it difficult to demonstrate *any* rule of post-marital residence. This is clearly not the case for the Gainj; a sharp difference between males and females exists for dispersal distances and events. Wood et al. (1985) state that males typically stay within one or two kilometers of their birthplaces for the duration of their lives. Females show a similar pattern until marriage (*c.* 20 years old) whereupon they move to the residence of their husbands. Thus, virtually all dispersal is of females moving at the time of their marriage and most of this movement is confined to within 10 kilometers of their birthplace. The fewer males who do disperse, travel farther than females (8.1 versus 6.3 km).

Wood *et al.* (1985) further explore the importance of two other variables in addition to endemicity (tendency to remain in the natal parish) and distance to Gainj–Kalam dispersal – language differences and population size differences. Although not as important as endemicity and distance, linguistic differences do structure the pattern of movement. When the differences between the Gainj- and Kalam-speaking parishes are considered, it is shown that male dispersal is strongly influenced by linguistic distance (and this correlation removes the effect of geographic distance *per se* on males) whereas female movement is affected by both linguistic and geographic distance. For males, this difference can be explained by the political importance of living in a parish where they speak the principal language. In these societies, males compete for prestige and speaking ability plays a role in mobilizing support.

Relative population sizes of the parishes also play a role in dispersal. Both males and females show negative density-dependent movement. That is, individuals tend to move to places with less dense populations. The high degree of philopatry is a result of people preferring to maintain ties with kin and access to garden land. Short range movement allows individuals to continue their social interaction with kin at home. However, if there is a shortage of land due to short-term excess of population in a neighboring parish, movement to a less crowded area would be advantageous. In contrast to the Yanomamo, the Gainj–Kalam populations are not expanding, apparently regulated by high density-dependent mortality (Wood & Smouse 1982) in relation to resources. Density-dependent migration may also reduce local population imbalances and consequently reduce mortality.

A further consequence of the Gainj stationary population relatively at equilibrium with respect to resource distribution is that the fission–fusion population structure found among the Yanomamo and the Semai is absent. The much higher population densities of the Gainj–Kalam leave no room for expansion. Fission groups would have no empty land to colonize nor would they be likely to be welcome in already densely settled parishes where large reserves of land would not be available for them. As Smouse and Long (1988:44) note, 'every parish fission or birth is balanced by a parish fusion or extinction'. Thus while some population readjustments

may be effected by marital migration reducing local imbalances in population concentration, the large scale population rearrangements typical of
fission–fusion events do not occur.

High density, intensive agriculturalists, local groups within state

Spain – Basque (1850–1910)

The Basques of the western Pyrenees region of France and Spain are
generally considered a linguistic and genetic enclave within the rest of
Europe. Their language is non-Indo-European, perhaps retained from a
period before the expansion of this language family. From the standpoint
of genetics, the global maximum frequency for the rhesus (Rh) negative
blood group allele among the Basque has invited the inference that they are
a unique population isolated from the rest of Europe. Indeed, based on the
expected outcome of selection at the Rh locus, it was postulated that the
Basque might represent an early European population that was originally
genetically uniform for the Rh negative allele (Boyd 1963; Levine 1963).

In other respects, however, Basque rural farmers seem not untypical
practitioners of the European system of mixed agriculture (Slicher van
Bath 1963). In contrast to the predominant focus on intensive rice cultivation in Asia, Europeans (including the Basque) combined stock-raising
with cereal crops. This mixed economy is able to support a relatively dense
human population. For instance, in the Bastán Valley of Navarra, Spain
during the late nineteenth century, Basque population density was approximately 25 persons per square kilometer (Abelson 1978). This density
is not that different from 'primitive' cultivators such as the Gainj living on
the fringe of Papua New Guinea's highlands and very much less than the
approximately 300 persons per square kilometer of India (see Table 2.1).

The Basque farming system combines an intensively utilized area for
cereals surrounding the farmstead (or in some areas, the nucleated group of
farmsteads or hamlet) with concentric zones of lesser intensity of usage at
greater distances used for pasture. The cereal lands are intensively cropped
with no fallow period, and elaborate cultivation techniques including
plowing and harrowing and manuring are employed. Additionally, in steep
terrain, the gradual slippage of soil must be countered by hauling it back
upslope by basket or cart, an extremely laborious task (Douglass 1975).
Contrasting with this pattern of intensive use of cereal lands are the zones
of meadowland, providing hay and green fodder for the animals, and the
even more distant, higher elevation forest and pasturage that supplies
wood, grass for animal grazing, and ferns for stable bedding. Less labor is

devoted to production on these lands and correspondingly much less output is realized. Thus, the Basque system combines a highly intensive cereal cropping regime with animal-oriented pasturage and woodlots, extending the acreage of the farm and thereby reducing the sustainable human population density when compared to systems devoted to grain production alone.

This description applies to the rural Basque population before circa 1925 (Douglass 1975). In recent times, Basque farming populations have been severely impacted by industrialization. Their region of Spain particularly has been the locus of new industries with consequences profound enough to have caused a 'rural exodus' (Douglass 1971) as farmers migrated to cities or overseas. However, migration is not only a recent response to changed economic conditions for the Basque, but is also a strategy of considerable antiquity. Basques were included in some of the earliest voyages to the Americas and have continued to migrate to both North and South America in considerable numbers (Douglass & Bilbao 1975).

The causes of this long-standing practice of emigration reside in Basque demography and economy. Most importantly, the Basque territory in Spain and France for a long time has been densely occupied in relation to the support capacity of their farming economy. Basque farmsteads are named and these names have persisted as entities over the centuries – every field and mountain tract is named and bounded by markers (Douglass 1975:42). This concern with territory may not imply saturation of the habitat at carrying capacity, but it does suggest that extensive lands are not available for additional people.

If the Basque population was stationary, as for the Gainj (Wood & Smouse 1982), the lack of surplus land would not cause a problem. However, given the relatively large Basque family sizes, each generation includes additional offspring with no expansion in the land base to provide for additional farms. This dilemma is resolved by an impartible inheritance system (also a feature of several other European populations, the Irish being perhaps the best known, Arensberg & Kimball 1968). Since the habitat has been essentially completely partitioned, each farmstead is no larger than required to support a family. Farm land and resources, therefore, must be maintained intact; thus, the farm house, land, stock, and tools are inherited as a unit by one of the children (either by primogeniture or 'suitability' as determined by the parent). 'Inheritance' usually occurs before the death of the previous head of household so that the occupants of a Basque farmstead may include multigeneration stem families comprised of the retired parents, one of their children (male or female) who has 'inherited' the estate, his/her spouse and young offspring, as well as any unmarried siblings of the inheritor who choose to stay and work the family farm (foregoing reproduction). The choice for non-inheriting siblings, then, is relegation to 'old boy or girl' status if they remain with the family, or migration, ideally with a dowry as their portion of the estate. This marriage system also seems a common European pattern (Goody 1976). As Douglass (1975:120) concludes, 'the Basque system of domestic group organization depends upon migration and emigration of excess membership'. Emigration outside the local area was necessary in order to earn a living.

The consequence of this inheritance/migration system is the extreme localization of one component of the population, the inheritors, and the exclusion from this local population of the disinherited. Thus, Borella (1994), who studied a Basque community in northern Navarra, found that local endogamy rates for individuals not emigrating was quite high. For land-owning inheritors, all marriages were between partners born within a 20 km radius of the community both for the period 1900–10 and continuing into 1950–60 even as other persons moved away. Similarly, Abelson (1976) found the mean marital distance in another Spanish Basque region to be only 10.4 km. Thus despite a tradition of emigration, often to distant countries, within the Basque homeland migration is not particularly long distance.

Viewed from the perspective of the Johnson and Earle (1987) taxonomy, the Basque practice a peasant economy within the nation-states of France and Spain. Rural Basque communities are part of a larger polity that is hierarchically structured. At the same time, Basque farm households are independent economic units. Basque reckon kinship bilaterally but the potentially wide network of ties implied by this system is not activated and there are few relationships between different domestic groups (Douglass 1975:34) – loyalty is to the self-sufficient farm family.

Kinship, however, is important in structuring long distance migration. Emigration to distant locales, especially the Americas, has been a continuing feature of Basque history. Initial destinations in the New World included both North and South America and the islands of the Caribbean. Once settlement was established, the kinship networks between previous emigrants and homeland potential migrants stimulated and guided further movement (Douglass 1975:125). One example noted by Douglass (1975) from the 1930s was the recruitment of a group of people from one village to migrate to Australia on a labor contract. The kin ties between villagers and their relatives in Australia created a channel for further migration.

Social stratification is a consequence of increasing economic intensification according to Johnson and Earle (1987). Although most Basque social relations are concentrated among domestic family groups, the society as a whole is to some degree stratified. The primary factor distinguishing class is ownership of land (Abelson 1978). Based on parish records from 1850–1910 and census data (1897) from two villages in the Baztán Valley of Navarra, Spain, Abelson (1978) found that owners occupied some 38 percent of houses and tenants rent the remaining 62 percent. The current ratio of owners to renters is essentially reversed in the Spanish Basque villages studied by Douglass (1975) – 29 percent are renters in one village, 37 percent in the other. Borella (1994) found for another area of Navarra in the 1900–10 era renters constituted 37 percent. Thus considerable variation in the ratio of owners to tenants has existed at different times and places.

Classes defined in this way do not seem to be immutable categories and in theory tenants could pass on the use of the land to their chosen heir (Abelson 1976; Douglass 1975). Nonetheless, this distinction had demonstrable effects on endogamy and migration (Abelson 1978; Borella 1994). Owners are more likely to marry locally and to spend their entire lives in the same house; tenants are much more mobile. Abelson (1978) found a significant difference between the owners and renters in mean parent–offspring birthplace distances. For owners, more than 70 percent of offspring were born in the same place as their fathers – a little more than 50 percent for renters. Consanguineous marriages were also higher among owners, 16.3, compared to 3.7 percent for renters. Borella (1994) found similar differences in the area she studied (70% owner endogamy for the 1900–10 population versus 48% for renters). She also points out that family history data supports the greater mobility of renters.

Turnbull (1981) made an interesting observation on the effects of the Basque inheritance pattern on the maintenance of the Rh blood group polymorphism. The high frequency of the Rh negative allele, d, is one of the most salient features of their presumed genetic uniqueness. Turnbull pointed out that the incompatibility selection associated with the Rh locus is birth order specific, that is, first born offspring are usually not selected against due to the lack of maternal antibody. Since primogeniture is a common basis of Basque inheritance, farm owners remaining in the local area would usually be first born sons or daughters whereas their later born siblings would emigrate. Therefore, the resident core Basque population would have experienced minimal selection due to maternal–fetal incompatibility.

England – Oxfordshire (1851)

A detailed study of a group of English villages in the Otmoor region of Oxfordshire was carried out by Harrison and his colleagues (Harrison

1995). This is a rural area of England comprising some 4000 acres with several villages surrounding a moor and settled since Anglo-Saxon times. One of the great advantages for this study was the availability of excellent historical demographic records maintained in the parish books of the Church of England. For some villages, these records of baptisms, burials, and marriages date back to the sixteenth century, and national census data also became available starting in 1801. Thus a much fuller historical view of population and mobility is available for this population than for many anthropological studies.

Over the some 300 years covered by the historical records, changes in various attributes of the population can be documented (Harrison 1995). The most profound modifications of Otmoor are recent, however, a major factor affecting movement was the coming of the railroad to the region around 1850. Another important change in the Otmoor region parallels a similar trend throughout the developed and developing worlds – the progressive decline in the importance of farming. Recalling that Ravenstein (1885) was promulgating his 'laws of migration' in industrializing England of the nineteenth century, it is to be expected that Otmoor also was experiencing some of the flows of rural people to the new urban centers. Further, the social class structure that has divided English society was already in place prior to recent modernization. Thus Otmoor, although rural and isolated by English standards, was hierarchically stratified into classes as well as spatially divided by geography and distance.

The relatively high population density (200 persons per square kilometer) of the region in 1851 reflects a nucleated society divided into small villages of a few hundred inhabitants which was also occupationally and socially divided. Interestingly, the population of the area (3142 in 1851) has declined over the subsequent decades and in 1961 was only about 2500 (Harrison 1995).

The Oxford study's dependence on parish records and census materials meant that some kinds of migration were less likely to have been ascertained. Parish records refer to the local residents and exogamous marriages are recorded elsewhere and were unavailable. Further, parish registers of marriage list the *residences* of bride and groom rather than their birthplaces. Harrison (1995) notes, however, that census data from the nineteenth century showed that most people were resident in their birth places at the time of their marriage. These data suggest that most permanent movement occurred as a result of marriage and the spousal residences at marriage reflect this movement.

The marriage records show a gradual increase in the rate of parish

44   *Anthropology of human migration*

exogamy over the period of the study from 1601 to 1951 from only about 20 percent to over 80 percent. When map distance is the metric, however, a major break occurs around 1850 with the coming of the railroad. Prior to 1850, the marital distance 'remained remarkably constant at 6–8 miles' (Harrison 1995:46), subsequently, it increased to around 30 miles (*c*. 48 km). Surprisingly, there has been no obvious rise in this later figure in more recent times. Coleman's (1977, 1979) data from other English regions corroborates this finding. Harrison suggests that increasing ease of travel may not translate into courtship at very long distances. Plots of marital distance for one village both prior to 1850 and after 1850 show a high frequency of mates found at close range (within a dozen miles). The main difference between the two distributions is the increasing frequency of longer distances, particularly in the range of 35 to 75 miles (*c*. 56–121 km). These distances suggest that rail transport opened up the possibility of meeting and marrying spouses from the growing centers of London and the industrializing Midlands. Even so, as Harrison emphasizes, the preponderance of marriages continued to be local.

Harrison and his colleagues (Boyce *et al.* 1967) have devised a model relating marital distances in populations such as Otmoor to visiting and the accumulation of knowledge about the local area. This argument will be examined in detail in Chapter 3.

English society is more complexly stratified than the Basque social division into 'owners' and 'renters' discussed in the last section. According to Harrison (1995), the spatially defined population of Oxfordshire is vertically stratified into a number of subpopulations based on class. In this sense, social classes are components of population structure.

Harrison (1995) identifies occupation as the principal criterion of class in Oxfordshire. He is able to obtain data on the occupations of the historical population from both national census and parish records (although he concedes that there are imprecisions in these sources) and from direct interviews with the contemporary population. These data allow separation into five classes from I (professional) to V (unskilled laborers); further subdivision into agricultural and non-agricultural work is made within the classes for some purposes. The distribution of members of the classes varies temporally and by size of parish, for example, numbers of agricultural workers have diminished through time and larger parishes have more service occupations.

A striking effect of class structure on movement is apparent in the Oxfordshire study. Taking marital distances as the metric, Harrison (1995) reports that the mean distance between residences of prospective spouses was 56 km for social class I men, 37 km for non-agricultural class II men compared to only 11 km for agricultural workers. These differences extend to post-marital movement as well.

Another aspect of interest is the degree of social mobility between classes through time. Social classes are subdivisions of the population into which individuals are born. Initially, at least, children are members of their parents' class. However, just as a person may move to a new geographic subdivision, people may move up or down in the class hierarchy between different social subdivisions. As Harrison (1995) notes, such movement can occur by marriage, often as women join their husband's occupationally based social class, or as individuals change occupations. Clearly if no such movement were possible, totally endogamous social classes could divide the population into separate gene pools.

Based on the Oxfordshire data, Harrison (1995) shows that social mobility is sufficient to homogenize the population genetically over a time span of less than 20 generations. Both marital exchange and individual social mobility combine to lead to high levels of relatedness between classes. Thus, however strongly entrenched the English class system may appear in the short-term, social subdivision does not appear to lead to sharply demarcated gene pools over the longer term.

Northern India – Faizabad District (1951)

Among the highest population densities in the non-industrialized world were found in village India. The northern state of Uttar Pradesh in 1951, for instance, comprised some 64 million persons in an area of circa 113,000 square miles (*c*. 190,970 km^2); within this state, the District of Faizabad studied by Gould (1960) included 1.5 million people in an area of 1740 square miles (a population density of roughly 300 persons per km^2). This very dense population, divided into an elaborate hierarchy of 'hereditary' occupational specializations (the caste system), was supported by intensive agriculture. Agricultural land was valuable and it, along with the 'property' associated with caste occupations, according to Gould (1960:480) was inelastic and 'firmly cements families to their community and local region'. Under such a regime, migration should be severely curtailed and of short range. However, two other aspects of northern Indian social organization affect marriage practices and consequently marital migration. These are caste endogamy and kin-group exogamy.

Each of the numerous castes into which northern Indian society is divided is in theory a closed breeding group. Mates must be obtained within the caste. Thus the potential mate pool for any rural Indian could be very much smaller than the census population. As Gould (1960) notes, this restriction is especially applicable in areas where the population composition has been stable. Since there is little mobility due to the tight ties binding families to their land, the members of a caste in a region are usually closely related and may even all be members of a single descent group. Even where these relationships are not genealogically traceable, the ideology of kinship for all members of a local caste is maintained. But to marry a relative is also prohibited; that is, agnatic kin groups (*gotras*) are exogamous (Gould 1960). This is particularly true of the high caste Brahman groups which are considered the most orthodox and to be emulated by other castes. The territorial localization of kin groups means that gotra exogamy is equivalent to village exogamy (at least in so far as the caste attempts to follow Brahmin practice).

The actual distribution of marriage distances and mobility is a compromise between the centripetal effects of territorial localization and the centrifugal effects of kin exogamy forcing marriages outward. As Gould (1960) shows, for most castes, the high density of people in the region he studied led to relatively short average marital distances (only 10.5 miles – c. 17 km – averaged over all castes for his main study village). Thus despite village exogamy, marriage is mainly local. Of the 345 marriages he recorded, 93 percent occurred in the western portion of the district. Interestingly, district boundaries seem also to be mating boundaries in this region of India – even where villages of other districts are geographically closer, marriages are less likely than with more distant intra-district villages. Within the district, nearby localities are most attractive but other considerations also apply. The network of roads influencing traffic flow is one such factor. Also significant are the prior ties established with a neighboring area. Gould (1960:482) notes that once intermarriage has occurred for 'whatever initial reasons', continued interaction is maintained 'because affinal ties that get established become a source of information concerning further marital opportunities', an obvious connection to the 'mean information field' model of Morrill & Pitts (1967) discussed in Chapter 1.

A further consequence of the Indian system is that higher castes differ significantly from lower in the distance from which they obtain mates. Gould divides the population into three caste groups: Group I includes the Brahmins and Rajputs, the principal owners of land; Group II comprise the several farming castes; and Group III includes landless artisans and laborers. Table 2.3 reproduces Gould's (1960:483) figures for marriage frequencies within and outside the western portion of the district for the three caste groups.

Although these percentages represent a rather small number of marriages (26 marriages for caste group I; 345 marriages in total), clear differen-

Table 2.3. *India: distribution of marriages in western Faizabad*

Caste group	Areas within western district			Outside western district
	1	3	3	
I	0.12	0.15	0.08	0.65
II	0.72	0.22	0.04	0.02
III	0.59	0.35	0.05	0.01

ces are apparent in these data. The Brahmins and Rajputs of Group I obtain mates from a much wider universe than do the other castes. Whereas 98 percent of marriages of Groups II and III are within the western district, 65 percent of Group I are outside; the average marriage distance for Group I castes is 40 miles (*c.* 64 km), usually outside the district.

The basis of this greater mating range of Rajputs and Brahmins lies in the political and economic integration of these groups over a wider area. Historical links of kinship and politics with other clans in other regions extend their spatial boundaries. Another reason for this greater range may be that supplied by Gould to explain the difference between Groups II and III. Group II, composed of cultivators, is the most populous grouping. The artisan–laborer castes (III) have fewer members in the local area and finding a suitable spouse is consequently more difficult. They must search further to find a mate and therefore show a lower frequency of close-by (Area I) marriages. Although Gould links the much greater spatial mobility of the Brahmins and Rajputs to status and political factors, it is also true that there are many fewer of these Class I caste members in the local area. Given chance variation in the numbers of potential mates as a result of demographic factors and marriage restrictions (kinship, spousal age-differences, etc.), a numerically small local mate pool may not include any appropriate person (MacCluer & Dyke 1976). In this northern Indian case, the average difference in mating distance between the more numerous Group II castes and the smaller Group III castes is quite small (6.4 miles or *c.* 10 km for Group II; 7.5 miles or *c.* 12 km for Group III). The very much greater distance for the Group I castes (nearly 40 miles) suggests that while small population size may play a role, Gould is correct to emphasize sociopolitical factors in the wider marriage universe of Brahmans and Rajputs.

Generalizations

This survey has compared a small sample of societies along a continuum of increasing population density, intensity of land use, and social integration.

No perfect correlation between position on this continuum and all aspects of migration was apparent, however, some generalizations emerge from these comparisons.

Population density, particularly at the low end of the scale, strongly conditions the mean marital distance. Where population densities are fewer than one person per square kilometer (e.g. !Kung, Aka, Vaupés), individuals marry within a much larger area than do spouses from more densely populated places. Clearly this relationship may be affected by the mode of transportation available, but when societies depend on foot or animal transport, equalizing the time cost of movement, a greater population concentration provides a sufficiently large potential spouse pool in a smaller area.

The higher travel cost of obtaining spouses from a greater distance may be required of members of low density populations because of the unavailability of potential spouses. There is good reason to believe that a minimum size is necessary in order for a population to be endogamous (MacCluer & Dyke 1976) (see Chapter 4 for the development of this theory). Below this minimum, demographic fluctuations in sex ratio, age and kinship structure reduce the probability of finding a suitable mate. As population density increases, local mate pools become larger and individuals need not travel as far to find a spouse.

Level of socio-cultural integration also influences mobility and migration. Societies in which the family is the highest level of integration are more likely to use mobility as a risk stabilizer. Recall that Johnson and Earle (1987) stated that greater political integration arises partly as a way to manage risk. Societies lacking such managers may depend on stored social obligations, visiting relations, and wide marital networks allowing access to resources over wide areas. Thus mean marital distances may be increased in such populations not only of necessity (increasing potential mate pools) but also to achieve economic and political ends. People marry far to extend the network of social cooperation and refuge in time of local shortage.

The lack of higher levels of political integration and authority in family-level societies also increases mobility since the ultimate resolution of dispute is to leave the group. Turnbull (1968) in fact saw this political factor as being the most important cause of flux in Pygmy local groups; fission and movement to another camp was a common feature of Pygmy life. Yellen and Harpending (1972) similarly described !Kung camps as 'anucleate' and implicated disputes as a factor in the continual fragmentation of these temporary assemblages. Semai and Yanomamo, although living in more stable local groups and possessing nominal 'headmen', also fission when the level of conflict can not be resolved in the local group. Indeed, the basic Semai response to *any* crisis was flight. Fission and fusion over a period of time reckoned in years and decades is characteristic of these groups as compared to the continual week to month flux in family-level societies.

With higher densities and more intensive use of land, mobility may be reduced. Land itself becomes a more valuable commodity and marriage and kin groups may be important regulators of access to land. Thus the strong commitment to philopatry of the Gainj (Smouse & Wood 1987) was explained as a need to retain ties with kin to be granted use of garden-land managed by the localized kin group. The marriage rules of mobile family-level societies requiring individuals to marry into distant groups to maximize the areal extent of kin are replaced by more localized alliances.

With regional systems and state organizations, 'order' is often imposed over large areas. Highly mobile populations seem to be regarded as antithetical to order and 'settling the nomads' seems an almost religious duty for state functionaries, a process continuing into the present as various governments strive to 'resettle' aborigines. Even the Great Wall of China may have been an attempt to keep the Chinese within bounds as much as it was to keep nomads out (Lattimore 1951), serving as a line demarcating the area of intensive agriculture integrated and controlled by the Chinese state and the steppe, an area of extreme dispersed nomadism uncontrollable by the state.

Within the boundaries of states, peasant economies may function on a very local level as illustrated by the Basque and Oxfordshire marriage patterns. Relatively dense settlement and intensive agriculture drive up the value of land and make inheritance an important strategy. The need to receive land and chattels from the family binds some members tightly to the local area. At the same time, the very scarcity of land insures that not all offspring can inherit. Therefore, an intensive, locally focused agrarian system may force some individuals completely outside the system. The Basque case illustrates this dual mobility pattern – localized inheritors continuing in the family estate and their non-inheriting siblings often dispersing great distances.

The differential in economic position and social status that has developed in state societies also affects marriage patterns and mobility. Among the Basque, owners (usually inheritors) of farmsteads are much less mobile than renters. The Oxfordshire population is more hierarchically arranged into classes and the caste system of India is an even more rigidly structured social hierarchy. In so far as these class and caste divisions follow rules of endogamy, local and regional populations may be stratified into separate breeding populations. This tendency exists among the Basque and the English, although as Harrison (1995) points out, social mobility is enough to retard genetic differentiation. Even the Indian system allows some mobility. Thus, over the short-term, however social stratification may affect marriage and mobility, it surely has not caused (nor is likely to ever produce) the division of populations into the Eloi and Morlocks envisioned by H. G. Wells in his classic fantasy *The Time Machine*.

Finally, perhaps the take-home message from these comparisons is that there is not one 'real' human population that typified human populations throughout the long span of our evolution. While it is highly unlikely that caste relations prevailed in the Palaeolithic, it is equally unlikely that all these populations were Yanomamo-like or even !Kung-like. Mesolithic Europeans were living under environmental conditions that were closer to the American Northwest Coast than to Australia and were more likely to have experienced demographic regimes, densities, and mobilities commensurate with these variables than with modern desert foragers. As Smouse and Long (1988) point out, over the long span of human prehistory, cycles of population expansion into new territories were followed by population growth and 'habitat packing' leading to more localized mate exchange. During the expansion phases, population structure and its genetic consequences might have approximated the Yanomamo pattern. Stable populations of foragers in sparse habitats might be expected to share basic features with the !Kung and, with increasing population densities, Gainj-like philopatry may have prevailed.

The key to applying our understanding of modern-day demographic and population structure variation to the evolution of past populations is the identification of archaeologically recognizable signatures for variables such as population density, intensity of land use, and degree of social integration. Granting the difficulty of achieving such estimates, even without exact knowledge of the values of these variables, ballpark estimates should allow much more realistic model building.

3 *Population genetics models and human migration*

Classic models

The classic population genetics models of migration emphasize generality and mathematical tractability rather than realism (Levins 1966). Sewall Wright (1931; 1969), who is especially responsible for developing much of this theory, aimed to produce a mathematical treatment applicable to all species. Such generality has great advantages and has served to unite botanists, zoologists, and anthropologists within a common theoretical enterprise. But generality has a cost; that is, a more precise understanding of population structure and evolution in particular species or populations will not be pursued. Further, the simplifying assumptions that allow general models to be formulated may often be violated in particular situations. Inferences made on the basis of models whose crucial assumptions are not met obviously will be flawed.

If the deviations from assumed model conditions are not great or the predictions need not be precise, the cost of simplification is slight. To illustrate: some years ago, Morton (1977) claimed that Malécot's (1955) equation relating genetic similarity and distance provided an 'acceptable' fit to many human populations. Morton also admitted that this very general model provided 'a less complete and reliable prediction' than a more detailed model. Thus, if it is sufficient to know that genetic similarity declines more sharply in continental 'isolates' than among hunter–gatherers or oceanic islanders (Morton 1972) and if geographic distance is the only variable considered relevant, then the model conveniently summarizes the relationship. However, when more careful attention is paid to the levels of subdivision of populations, it can be shown that geographic distance is not the only variable determining genetic similarity and its effect varies with the level of population subdivision (Fix 1979).

One of the great advantages of the study of human populations is the wealth of information available on their social and demographic structures. Along with these data on the current state of the population, genealogical and historical records can provide a temporal dimension, information that is rarely available for natural populations of other species. Human populations, therefore, offer a better opportunity to assess the assumptions and measure variables for the mathematical models of population genetics than do populations of many other organisms (Fix 1979). With this potential in mind, the following sections will specify the key assumptions of the classical models of population structure as they relate to migration.

The island model

The basic population structure underlying much of Sewall Wright's mathematical theory is the island model (e.g., Wright 1951;1969). Many of the predictions concerning the relative importance of genetic drift, natural selection, and mutation are based on inequalities derived from this basic model. Crucial formulae specifying the expected degree of inbreeding in subpopulations also rest on the assumptions of the island pattern. For example, consider the often reproduced equation specifying the equilibrium between migration and genetic drift,

$$f = 1/4N_e m + 1$$

where f is the inbreeding coefficient in the subpopulations, N_e is the effective subpopulation size, and m is the migration rate. From this simple equation comes the prediction that if m is much smaller than $1/4N_e$, then f is large and there is a high degree of local inbreeding and homozygosity, whereas the opposite condition leads to reduced inbreeding and ultimately to panmixis among all subpopulations as a single large breeding population. A commonly cited result of this equation is that a very low migration rate, on the order of one migrant per generation, is needed to conteract local differentiation (Crow & Kimura 1970; Spieth 1974). The ramifications of this finding are extensive, in particular, it would imply that genetic microdifferentiation could only occur under conditions of almost complete subpopulation isolation. However, in the island model, migrants are assumed to have a gene frequency equal to that of the entire assemblage of subpopulations. Note that this frequency is not a random variable, migrants are not a *sample* of the entire population and therefore the gene frequency of the migrants has no variance.

Gene frequency change in each subpopulation, then, where only migration is considered, will be a function of gene frequency in the subpopulation, p_i, m, the migration rate (specifically, the proportion of the subpopulation replaced by migrants each generation), and p, the mean gene frequency of the entire population. That is:

$$\Delta p = -mp_i + mp$$

The stabilizing power of migration is obviously very great in such a system. Each generation local gene frequencies are pushed toward the overall population mean gene frequency by migration. As m approaches 100%, each subpopulation gene frequency, p_i, will become the population mean frequency, p. Further, m will always decrease local genetic differentiation due to processes such as genetic drift or localized selection.

This model of migration in a system of subpopulations is wonderfully simple mathematically, it allows the elaboration of combinations of variables in Wright's comprehensive theory (1969). However, it is hard to imagine an actual empirical situation corresponding to the model. Potentially, a single island near a large panmictic mainland population such that migrants to the island might represent an equilibrium population (too large to be affected by genetic drift) might qualify. But, as Crow and Kimura (1970) state, migrants are more likely to have come from neighboring subpopulations that are likely to be more alike genetically than more distant subpopulations (the principle behind the concept of isolation by distance to be considered below). The usual sense of a species or regional subdivided population entails greater geographic proximity between some subpopulations and isolation by distance seems inevitable. Thus Wright's model is highly unrealistic. The question remains whether this lack of realism matters when general answers to broad questions are sought.

Isolation by distance

At the opposite end of a continuum of population structures from the 'isolated' island model is the continuously distributed uniform population implied by Wright's isolation by distance model (Wright 1943; IBD). There is no analog in the IBD model to the randomly mating 'islands' receiving migrants from an infinitely large panmictic source. The population is not subdivided into demes that exchange migrants or receive migrants nor is it a panmictic unit itself. Random mating is limited by distance such that individuals are more likely to encounter and mate with neighbors than with those farther away. Groups of individuals may thus be clustered into 'neighborhoods', areas defined by 'central individuals' whose parents may be treated as if drawn at random (Wright 1969:295). Genetic variation within the population will depend on the size of these neighborhoods, which is also to say that the mobility of individuals or the distance from which they choose mates, defines the spatial structure of genetic variation.

The simplest representation of IBD is a linear (one-dimensional) habitat along the length of which the difference between parents and offspring birthplaces are normally distributed from one generation to the next (Wright 1969:297). Neighborhood size is a function of the standard deviation of the parent–offspring difference. More complicated mathematically is the two-dimensional case of a uniform area of individuals; conceptually, however, the model is identical. The effective population size in a uniform area is equal to $2\pi\sigma^2$, where σ is standard deviation of parent–offspring birthplaces. Wright (1969) also considered the case when the distribution of parent–offspring birthplaces is not normal, especially for the often observed situation where most mates (and resulting offspring) are found at close distance.

While perhaps more realistic than the island model, the assumptions of IBD are probably never met in real human populations. Regional populations of hunter–gatherers such as the !Kung considered in Chapter 2 are so fluid that they may approximate a continuous distribution of individuals. Similarly, areas of dense populations of intensive farmers undivided by linguistic or political boundaries might also nearly satisfy the assumptions for IBD. However, as documented in Chapter 2, populations of state-level societies are often hierarchically stratified and marriage is not random within areas. The northern Indian caste system most clearly demonstrates violation of panmixis due to social barriers. Again, the IBD model, with its strict simplifying assumptions, allows the development of mathematical theory to explore the long-term consequences of a generalized pattern. As always, the question is how critical is the violation of these assumptions to the particular problem being investigated. When general answers are sought, general models may provide them. The mistake is to apply these models to situations for which they were never intended.

Malécot's isolation by distance model

Malécot's (1950; 1973) treatment of IBD is the continuous population model most widely applied to humans. Cavalli-Sforza (1984) suggests that the popularity of this approach is due to its mathematical simplicity compared to the Wright approach. Certainly Morton (e.g., 1973; 1977) and his colleagues essentially redefined the concept of 'population structure' as the fitting of the Malécot equation to human data. Numerous populations ranging from hunter–gatherers (Australian aborigines) to modern state societies were studied using variously genetic, anthropometric, pedigree, and migration data (see Jorde, 1980:171–3 for a list of these populations). Not surprisingly, all this activity generated some controversy and a variety of critiques of this approach now exist (Cannings & Cavalli-Sforza 1973; Cavalli-Sforza 1984; Felsenstein 1975; Fix 1979; summarized in Jorde 1980). Rather than reiterate all this discussion, consideration of this model will be limited to its suitability for analysis of human patterns of migration.

Malécot's model assumes a continuous uniform population with migration simply being a function of geographic distance. Genetic similarity, defined as a coefficient of kinship ϕ, declines with distance following a negative exponential function, i.e.,

$$\phi(d) = ae^{-bd}$$

The parameters of this model, a and b, defined as 'local kinship' and 'rate of decline' respectively, are estimated by

$$a \cong 1/(1 + 4N_e\sigma m^{1/2})$$

where N_e is the effective population size, σ is the standard deviation of marital migration distances, and m is called 'systematic pressure', and

$$b = (8m)^{1/2}/\sigma$$

although these quantities are actually estimated by regression from the data in empirical studies (Jorde 1980). Later additions to the model included adjustments for the dimensionality of the habitat (although again these were little used in actual applications) and the addition of a correction factor, L, to compensate for the fact that ϕ, a similarity measure, was often negative for long distances (see Jorde 1980).

The basic similarity of this model to Wright's IBD is apparent – both see genetic similarity developing in local neighborhoods as a function of the degree of dispersal. Wright's basic model considers the distribution of parent–offspring birthplace differences to be normal while Malécot treats marital distances as a negative exponential distribution. Wright defined genetic similarity as a correlation; Malécot as a probability or kinship coefficient. For individuals, the kinship coefficient is defined as the probability that two random homologous genes will be identical (Crow & Kimura 1970:68). The difference parallels different definitions of the inbreeding coefficient as the correlation of uniting gametes versus the probability of identity by descent (Jacquard 1975).

Malécot's formulation also shares the emphasis on generality and simplicity of Wright's models. Surely geographic distance is an important determinant of human movement but, as has been documented in previous chapters, it is not the only variable that structures human movement. Malécot's model further reduces this relationship to a linear, one-dimensional case (Cavalli-Sforza 1984), a highly unrealistic assumption since most human populations exchange mates with neighbors on all sides. The model also reduces the distribution of marriage distances in all human populations to the negative binomial (see Swedlund, 1980 for further discussion of the assumptions of this model).

More problematic than these simplifying assumptions, however, is the parameter, m, the 'systematic pressure'. Although perhaps unrealistic, the exponential distribution of mating distance is at least a clearly stated and easily comprehended assumption. In contrast, m conflates several potential evolutionary processes including mutation, stabilizing selection, and 'long-distance' migration. Leaving aside mutation, which as Cavalli-Sforza (1984) notes, may be negligible in its effects on genetic similarity at this scale, and selection, which for particular alleles may have crucial effects on any geographic scale, the concept of long range migration as a stabilizing force has serious problems. The ability of this form of migration to over-come the random differentiation of local regions follows directly from the assumption of the island model that migrant gene frequencies are an average of the overall population and constant through time. This assump-tion exists in the absence of any actual data on patterns of long-range human migration or the gene frequencies of such migrants. It is certainly possible to argue that a small number of random migrants from distant locales might represent highly *unrepresentative* samples of their parent populations. Such migration might in this way contribute an overall stochastic rather than stabilizing effect to the population system. Just as the founder effect is seen as a form of random genetic drift, migrant groups may also constitute a random evolutionary process (see below for the additional bias caused by kin-structuring of migration). In any case, if populations are truly distributed continuously with no discontinuities between neighboring groups, then the distinction between local and long-distance dispersion is simply arbitrary (Cavalli-Sforza 1984).

The application of Malécot's IBD model has provided a very general view of the pattern of human genetic variation and, as Cavalli-Sforza (1984) and Morton (1977) have indicated, the picture is consistent with common sense expectations. Thinly distributed populations such as Australian de-sert aborigines show a slow decline of similarity with distance; continental 'isolates' show a sharper decline. But, as Cavalli-Sforza (1984) also states, to describe human population structure using a model of only two par-ameters is oversimplification. Particularly since human data allows us to evaluate much more detailed models, it seems a mistake to limit its analysis to simple distance models.

Case study – Finland

The analysis of the genetic structure of Finland (Workman *et al.* 1976) provides a nice illustration of the value of a comprehensive study of the role of distance at different levels of population hierarchy. The advantage of this study was that a large random sample of the population from the entire country was obtained. This wide sample allowed partition into eight counties and 27 geographic districts allowing IBD to be analyzed at a fine geographic grain and also at the coarser grain represented by pooled individuals in counties.

Recall that the Malécot model assumes that genetic similarity, ϕ, decreases with distance as a function of gene dispersal through marital migration. It might be expected that this effect would be most pronounced over relatively short distances. As distance increases, other factors might come into play obscuring the particular relationship of marriage dispersal and distance. Thus the purported cause of a decline in genetic similarity with distance, marital movement, may become a catch-all explanation for whatever is observed.

The analysis of genetic variation among Finnish counties is consistent with ethnohistory. At a finer level of population subdivision, the districts again show agreement with expectations from known history. When pooled into distance classes, the districts also show a classic decline of genetic similarity (Figure 3.1). Following the practice of the Morton school (Morton 1973), the Finnish population could be added to the list of human groups for which an 'acceptable' fit has been achieved. However, when the data are disaggregated, it turns out that 'most of the effect is due to the geographic and genetic regional substructure . . . and not to any systematic effect of distance' (Workman *et al.* 1976: 358). Figure 3.2 shows the relationship between genetic similarity and distance for three of the districts (where each is compared to all other districts).

Districts within the same region are labeled in Figure 3.2 (SW: southwest, filled circle; C: central, open circle; NE: northeast, filled triangle). These three diagrams represent some of the heterogeneity in patterns among districts. District 7 shows some decline of genetic covariation (r_{ij}) with distance, but the highest values are all with other SW districts (1, 3, 5); District 27 shows no effect of distance within its own region or with those outside the NE. Workman *et al.* (1976) provide other examples and should be consulted for further details. These data demonstrate that the apparent decline of genetic similarity with distance is an artifact of pooling rather than an effect of distance *per se*.

The more general point to make from this analysis is that the underlying

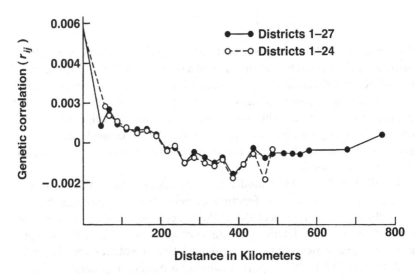

Figure 3.1. Isolation by distance in Finland. Genetic correlation (r_{ij}) with distance for all pairs of districts pooled into distance classes. (From Workman *et al.* 1976, fig. 6.)

genetic process responsible for a genetic distribution can not unequivocally be inferred from the genetic structure (Workman *et al.* 1976). To conclude from Figure 3.1 that a Malécot process of IBD is responsible for the observed pattern of decline in genetic similarity would be wrong. The availability of other sorts of data may help resolve this quandary. Thus the lack of effect of distance at longer than 50 km ranges does not obviate the very local effect of distance. A detailed archival study of seven Finnish communities between 1790 and 1915 showed 76.3 percent of all marriages were between mates born less than 10 km apart and only 1.1 percent more than 25 km distant (Workman *et al.* 1976). Isolation by distance should be pronounced over the short range, however, when distance classes of hundreds of kilometers are employed, genetic correlation or covariation is more likely due to some other cause than marital dispersion. In so far as the application of the Malécot equation presumes that marital migration is the principal cause of the genetic distribution, it is a misleading approach to take for all population partitions and all distances. As Workman *et al.* (1976) point out, Malécot's definition of kinship (or genetic similarity) depends on a specific suite of factors that bear on mating distances as well as the pattern of pre-existing kinship in the founder population and the time since foundation of the population system. Population density, a key parameter of the population classification in Chapter 2, strongly condi-

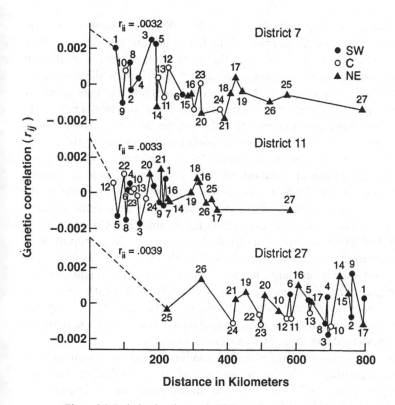

Figure 3.2. Isolation by distance in Finland. Genetic correlation (r_{ij}) with distance for selected districts (Districts 7, 11, and 27) compared with all other districts. Key – SW: southwest; C: central; NE: northeast. (From Workman et al. 1976, fig. 7.)

tions mating patterns and provides one indication of the level at which IBD should be important. Actual data on marriage and migration through time could more definitely identify the role of IBD. In the absence of such corroborating information, causal attribution based on genetic distribution alone should be unconvincing.

The stepping-stone model

The metaphor of stepping stones nicely describes the pattern of discrete subdivisions connected by migration that characterizes this model (Kimura & Weiss 1964). Each subpopulation exists either in a one-dimensional array or a two-dimensional lattice and exchanges migrants with its nearest neighbors (two in the linear case; four in the plain case). In contrast the previously described discrete model, Wright's island model, the stepping-stone model incorporates the IBD criterion seen in the continuous models. The subdivision of the population system into discrete stepping-stones, however, may more closely approximate the colonies or local demes of many natural populations including humans (Crow & Kimura 1970). Although more realistic in this sense, this model retains the basic simplifying assumptions of the other mathematical models of population structure: infinite number of colonies (corresponding to the infinite extent of the continuous population in IBD models); and, symmetric and isotropic migration of constant intensity persisting to produce genetic equilibrium (see Jorde 1980). Also shared with other general models is the notion of a systematic stabilizing force that might be mutation and/or selection but is usually described as long-range migration (Crow & Kimura 1970). Just as for the island model, each colony receives a proportion of migrants from the combined gene pool of the entire system. Since this metapopulation is very large, it is not subject to genetic drift and therefore the migrant gene frequency remains constant from generation to generation. This convention makes each colony of the stepping-stone system equivalent to an island and, in the case of no local migration, the stepping-stone model reduces formally to a series of independent islands.

The stepping-stone model makes similar predictions to other IBD models regarding the decline of genetic similarity with distance (not surprising since the model may be made equivalent to the continuous IBD model by decreasing the inter-colony distance to zero (Jorde 1980). The principal use of the model has been theoretical in contrast to the wide application of the Malécot model to empirical human populations. As with most theoretical applications, predictions concerning the pattern of genetic similarity or the decay of heterozygosity depend on constancy of conditions until an equilibrium is reached. Experience with actual human populations make these assumptions extremely problematic. Population sizes and migration rates often vary through time and are conditioned by location in space (non-isotropic). Depending on the rapidity of convergence to equilibrium, conditions may change drastically in a human population system, especially for groups such as the Yanomamo or Semai (see Chapter 2). Thus, when Kimura & Maruyama (1971) tell us that the condition for genetic microdifferentiation in the stepping-stone model is for $N_e m < 1$, it is not totally clear how applicable this result is for human populations, many of which may fail to meet the equilibrium assumption.

Migration matrix model

The migration matrix approach (Bodmer & Cavalli-Sforza 1968; 1974) makes use of the actual data on human migration in a region. The method takes the currently observed migration rates between each sub-population and uses them to predict the pattern of genetic variation at equilibrium. This avoids the onerous assumption of equal, isotropic migration rates and infinite number of subpopulations made by the stepping-stone model, thereby increasing the realism of the prediction. The trade-off for the advantage of greater realism is loss of generality. The predictions apply to the system as it was observed. Further, current migration rates between subdivisions may represent only a transient moment in the migration history of a system. To depend on today's rates to predict equilibrium genetic conditions that may take many generations to attain may be as unrealistic as the more general models. The migration matrix model also shares with the other models the reliance on 'systematic' pressure to stabilize the system. As for the other models, this force is usually attributed to long-range migration, another dubious assumption.

Jorde (1980) provides a condensed, clear exposition of several matrix methods including the Bodmer–Cavalli-Sforza model and that of Imaizumi *et al.* (1970) and his review along with the original sources may be consulted for details and formulae.

Since the matrix model does not depend on a general form for the migration pattern (e.g., the negative exponential function of the Malécot model), it would seem to have more versatility in dealing with the diversity of migration systems found in different human populations. By focusing on the actual rates of migration (usually parent–offspring birthplace differentials but sometimes also matrimonial distance data), the spatial details of different migration patterns are not submerged and more complex relationships between distance and movement can be investigated. In order to illustrate both the advantages and problems of this method, data from the Semai Senoi (Fix & Lie-Injo 1975) will be used.

Case study – Semai

The basic features of Semai population structure have been provided in Chapter 2. Recall that the Semai were (in 1968–69 when these data were collected), a forest swidden farming population living in small settlements of usually less than 100 inhabitants separated by at least two hour's walk. Censuses and genealogies were collected in seven settlements ranging in size from 50 to 272 persons. Parent–offspring birthplace data was obtained for most of the population allowing for the construction of a migration matrix.

Table 3.1 reproduces the backward matrix for the seven settlements based on some 1033 parent–offspring pairs. The elements of the matrix are derived from the observed numbers of individuals born in the ith settlement whose parent was born in the jth settlement. These observations include the entire population spanning three and sometimes four generations (following Ward & Neel 1970, and in contrast to the original Bodmer and Cavalli-Sforza one-generation model). Further, both parents are combined in these data rather than separated into father–offspring and mother–offspring matrices. Where there is a strong gender specific dispersal pattern (e.g., as for the Gainj considered in Chapter 2), separate treatment would be warranted. For the Semai, there is no preference for patri- or matrilocality and both males and females may be found in settlements other than their natal group.

The observed parent–offspring numbers are transformed into probabilities. Thus each element of the top portion of the matrix $(m \cdot j)$ represents the probability off an offspring born in settlement i having a parent born in settlement j. The upper portion of the figure shows migration among the seven settlements included in the study area; the lower portion separately details migration into the study area from settlements from other regions $(\alpha \cdot j)$. This convention (again following Ward & Neel 1970) subdivides the 'long-distance migration' component, α, of Bodmer and Cavalli-Sforza into contributions to the regional gene pool from specific localities and allows analysis of its constitution rather than lumping all 'outside' migration into a single category.

Figure 3.3 shows the spatial pattern of migration within the study area and into the area from surrounding areas. The intensity of movement is represented by the width of lines connecting the settlements.

The advantage of this method is that the actual pattern of migration, as measured by parent–offspring birthplace differences, within this area of Semai land is available for comparison with genetic measures (Fix & Lie-Injo 1975) and potential causes of movement. As will be seen, this *descriptive* value may not be matched by the *analytic* power of the method to describe long-term distributions of genetic similarity among subpopulations. The Semai, indeed, may be a particularly inappropriate population to meet the requirements of the analytic model. The detailed migration data demanded by the model, however, allow the examination of the fit between assumption and fact, a situation lacking in some of the other more generalized models such as the Malécot IBD equation.

Table 3.1. *Backward stochastic migration matrix – seven Semai settlements*[1]

		m·j						
Settlement	N	SA	KE	RU	KL	BU	KA	CH
SA	449	0.601		0.065	0.009		0.011	0.189
KE	82	0.085	0.171	0.049	0.146			0.159
RU	114			0.298	0.018	0.026		0.061
KL	70				0.430	0.200		0.057
BU	139	0.007		0.029	0.208	0.575		0.029
KA	26	0.039					0.115	0.403
CH	153	0.105		0.007	0.007	0.007	0.019	0.705

	α·j									
Settlement	001	002	003	004	005	006	007	008	009	010
SA	0.036	0.011	0.002	0.014	0.004	0.042	0.016			
KE	0.305					0.036	0.049			
RU	0.080	0.035		0.026		0.290	0.158			
KL						0.300		0.013		
BU	0.022					0.065		0.065		
KA				0.269	0.115				0.039	
CH	0.007	0.014	0.007	0.007	0.007	0.085	0.019			0.007

[1]Upper portion (m·j) of the matrix shows the probabilities of an offspring born in a row settlement having a parent born in a column settlement; the lower portion (α·j) shows corresponding probabilities for parents born in settlements outside the study area (both extant and abandoned).

Looking now at the Semai pattern, perhaps the most salient feature is the high percentage of children whose parents were born in another settlement. In only three of the settlements (SA, BU, and CH) were more than 50 percent of parent–offspring pairs born within the settlement. Even in SA, the largest of the settlements (population of 272 in 1969), nearly 40 percent of offspring had a parent born elsewhere. The two smallest settlements, KE and KA, show less than 20 percent concordance. Even the reasonably (by Semai standards) large settlement of RU (1969 population, 107) had less than 30 percent of shared birthplaces. To put these figures in perspective, consider the expectation for combined mother–offspring and father–offspring matrices under a system of mandatory exogamy. Assuming that one partner always stayed in their natal settlement and always married outside, the joint parent–offspring matrix should be 0.500 along the diagonal since for each child in each place, one parent would be local and the other from another settlement. Although the Semai do not specify settlement exogamy as a prescription (indeed, they seem to prefer, all other things being equal,

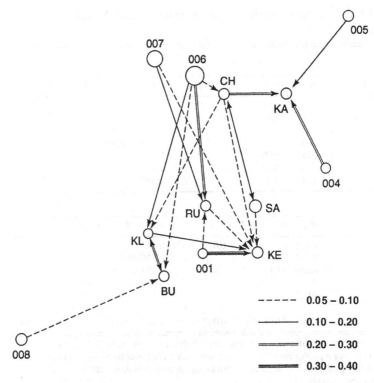

Figure 3.3. Semai migration pattern among settlements based on
parent–offspring birthplaces. (Data from Table 3.1; settlement labels also as in
Table 3.1.)

to find a mate locally – Fix 1982b) the actual distribution of birthplaces
suggests a great deal of out marriage.

Despite this apparent low degree of endogamy (see Table 2.2 for birthplaces of married pairs), mating and the consequent dispersion of offspring
birthplaces is not random. This is not a single panmictic population
considered over the three generation time span of these data. There are
definite spatial and social patterns structuring movement in this area.
These patterns are not well captured by the migration matrix formulation
(at least by that presented here with present-day settlements as the exchanging populations). The primary problem for analyzing Semai migration through a matrix approach is the instability of these local settlement
populations. A second problem is the decidedly local nature of 'long-
distance' migration. Discussion of both of these issues can help to suggest
when and where a migration matrix analysis might be appropriate.

As pointed out by Bodmer and Cavalli-Sforza in their 1968 formulation, real populations of humans are usually very irregularly distributed in space. Semai settlements cluster along river drainages of which there are no shortage along the flanks of the Main Range, the backbone of the Malaysian Peninsula and a major feature of Semai country. A major river system in the study area encompasses the settlements of SA, KE, RU, KL, and BU preceding from east to west and upriver. Travel along river drainages is usually easier than over the ridges and through the forest separating valleys (although note that settlement 008, outside the study area, is actually on the opposite flank of the Main Range and thus entails movement over the mountains). CH and KA are part of a more northerly system of rivers. The headwaters, however, of the tributaries of these systems (such as those along which RU and SA are located) nearly connect and there are well defined trails leading in these directions. Another point that is not obvious from Figure 3.3 is that this area is the southeast corner of Semai country. Nearly all other Semai live to the north or west; east and south are a few small aboriginal groups (Jah Hut and Che Wong), but mostly these areas are occupied by Malays. For historical and religious reasons, there has been practically no intermarriage between Malays (who are Muslims) and Semai (who are not). Thus socio-cultural 'irregularities' also structure the landscape and affect migration.

As important as the topography and geographic distance for understanding the observed distribution of parent–offspring birthplaces is the social dimension of kinship and historical ties among settlements. Again, migration matrices, by requiring actual data on dispersal, allow comparison of migration and genealogy and are more useful for this purpose than more abstract formulations.

The pattern represented by these data reflect the fission–fusion structure of Semai society (Fix, 1975 and Chapter 2). Thus SA, the largest settlement and one of the two most endogamous, is a long established place as is CH. Along with this relatively high degree of philopatry, there is a large amount of reciprocal exchange between these two groups. Some 20 years before the study, a large contingent split from CH and fused with the SA population. Given the several generations represented in the matrix, a high percentage of parents born in CH whose children were born in SA were members of this group. The connections of kinship and friendship formed by this fusion also promote continuing visiting and intermarriage between individuals in the two settlements. The relatively great geographic distance between SA and CH and their location in two different river drainages is less important than the shared history and kinship in influencing mating choice. BU and KL are spatially adjacent and are closely intermarried. Indeed, if considered a single population, their endogamy rates would exceed SA and CH. However, both have received significant increments of their populations from different sources outside the study area due to splinter groups fusing with them in the past. BU is also unique in the area for maintaining ties across the Main Range and many BU people claim kin in these settlements. Thus both RU and SA are in the same major river valley as BU and both have relatively large populations but neither contributes as many parents to the BU gene pool as 008 from across the Main Range. The present settlement of RU occupies an area settled long ago but was augmented in the recent past by a large group from outside the study area. Genealogies of older people demonstrate that RU formerly had closer ties with its neighbor, SA, which has been swamped by more recent movements of people from the north (especially 006 and 007). KA and KE are both relatively small and recently founded settlements. KE originated some 15 or 20 years before the study by the dissolution of a large settlement (001) a short distance to the west. From reports, a severe disease outbreak caused people to 'run away'. The core group of founders at KE were one component of this former population at 001. People with kinship ties to CH had lived at KA until about 15 years before the study. A large migrant group from the east (004), now forms the principal nucleus of the present population.

This brief synopsis of the history of fission and fusion, kinship links and marriage ties within this small area of the Semai distribution illustrates some of the factors that determine parent–offspring birthplace distributions among this population. This case also exemplifies a fundamental difficulty with the application of the migration matrix model. A basic assumption of the method is that the migration rates derived from current populations will remain constant over sufficient time for a genetic equilibrium to be attained (Bodmer & Cavalli-Sforza 1968). The number of generations required depends on the situation and the stringency of the convergence to equilibrium criterion (Jorde 1980). As the Semai case makes clear, however, current migration rates are unlikely to persist for even one or two generations as settlements fission and fuse or found new groups. Settlements as long persisting demes exchanging migrants at relatively constant and stable rates are simply not features of Semai society. Similar fission–fusion populations such as the Yanomamo (Ward & Neel 1970) also experience radical shifts in population distribution and migration flows over relatively short time periods. But, as Friedlaender (1975) pointed out for the highly localized Bougainville islanders (an Oceanic group similar to the Gainj described in Chapter 2), 50 generations of the same mating pattern 'pushes reality'. As suggested in Chapter 2, for populations such as the Yanomamo and Semai and certainly for the low density, mobile hunter–gatherers, the equation of village or camp to deme is incorrect, for even over the short-term such aggregations may be ephemeral.

The other general point to make from the Semai case is that the treatment of 'outside' migration as a systematic stabilizing force is highly problematic. Examination of the lower portion of Table 3.1 shows that migration into the study area has come from ten settlements not all of which are currently inhabited. The abandonment of settlement 001, for instance, provided some 30 percent of the parents of the current population of KE and there is some justification for considering these people to be 'residents' of KE. Although some of these locations (006 and 007, for example) probably include more than one settlement, none are large enough to be considered an unvarying gene pool. Some few cases of true 'long-distance' migration from other ethnic groups do occur in genealogies and some of these events seem to have had important effects on the Semai gene pool including the likely introduction of the malarial-adapted ovalocytosis allele (Fix 1995). These rare and essentially random events can not be considered stabilizing processes, indeed, they insert another form of stochasticity to the system (see Chagnon *et al.* 1970, for another example).

Consideration of the Semai case suggests that the migration matrix model, requiring long-term relatively stable demes exchanging migrants at a constant rate, is not universally applicable to human populations. The specification of current migration rates among settlements provided a useful descriptive framework for considering the causes and consequences of movement but the evidence shows that these rates would not persist long enough to reach an equilibrium. For more stable local populations found in intensive agrarian societies (such as the European and northern Indian societies described in Chapter 2), the assumptions are more likely to be met. Thus Cavalli-Sforza's (1969) original study in the Parma Valley of Italy comparing long-established village populations would likely be a suitable candidate for migration matrix analysis. Using a different matrix approach, Hiorns *et al.* (1969) showed that rapid convergence to 95 percent common ancestry occurred using the migration rates from their Oxfordshire study (see Chapter 2 and the next section). The point is not that migration matrix analysis is deeply flawed, but that careful attention must be paid to assumptions when choosing an appropriate model for the population(s) and problem(s) being studied.

Neighborhood knowledge model

This model is perhaps the only explicitly behavioral migration model proposed by human biologists. All of the previously examined models have looked at the consequences of migration recorded in the distribution of birthplaces or marriage distances which are then related to effects on genetic structure. Boyce and his colleagues (1967), however, based their model of marital migration distances directly on what they see as widespread characteristics of human behavior. Their exemplar population is a sedentary agricultural community whose members travel to surrounding areas during the course of their normal activities. This travel results in an accumulation of information about these neighboring settlements; i.e., *neighborhood knowledge*. As a result of the limitations of human locomotion by foot or by domesticated animal transport through most of human history, this knowledge will be constrained by distance – closer areas will be much better known than more distant locales. In a manner analogous to the mean information field model (Morrill & Pitts 1967) discussed in Chapter 1, Boyce *et al.* (1967) postulate that this knowledge determines the likelihood of choosing a mate from any population in the area. The distribution of marriage distances will therefore reflect the distribution of neighborhood knowledge and, as knowledge is reduced rapidly with distance, so will marriage frequencies.

More concretely, the model is based on the following explicit assumptions:

(1) The frequency of marriages with individuals from a neighboring village is directly proportional to the size of that village and to the frequency of visits to that village.

(2) 'The frequency of visits to a village at a particular distance from the home base is equal to the frequency of visits to all villages at that distance divided by the number of villages at that distance' (Boyce *et al.* 1967: 335). Included in this assumption is a further simplification, i.e., that villages are homogeneously distributed.

(3) Villages at a particular distance will be visited twice as frequently as villages at a greater distance since they will be passed through in transit to the more distant locales.

(4) The frequency of journeys to villages at a particular distance is inversely proportional to a power of twice that distance (Fix 1974).

Their model thus aims to explain the commonly observed exponential decline of marriage frequency with distance based on the size of home bases and the distance between them.

The form of the equation satisfying the model is a negative power function,

$$y = ax^{-b}$$

where y is mating frequency (measured in mates per 100 inhabitants) and x is distance. Given the assumptions of the model, the value of b should be close to 2; a is a function of the scale of distances over which marrriage occurs.

As Cavalli-Sforza & Hewlett (1982) note, this equation, often called a Pareto function, is widely used in geography and the Boyce *et al.* equation is not very different formally from the gravitational models used to model migration and mating distance (Cavalli-Sforza 1963). Another point to make is that the underlying rationale for this relationship is derived from the *concept* of neighborhood knowledge but, in the original presentation of the model, Boyce *et al.* (1967) made no attempt to measure this information directly.

Case study – Oxfordshire

The data that illustrated the model were derived from the 1861 census of the Oxfordshire parish of Charlton-on-Otmoor. The basic features of this population are described in Chapter 2. It will be recalled that this was a densely settled rural area of England with most marriages occurring locally (mean marital distance, 9.6 to 12.8 km).

Figure 3.4 reproduces the distribution of mating distances for the Charlton population. Each point indicates a village in the neighborhood of Charlton plotted with respect to the number of mates (per 100 population) that it provided to the 1861 inhabitants of Charlton and its distance from Charlton. Six miles (c. 10 km) was considered a reasonable limit to neighborhood size given the means of transportation available to this population. The high density of settlement in this area is apparent from the fact that some 23 villages are found within this small radius.

The continuous line in Figure 3.4 shows the fitted curve based on the model, $y = 4.75x^{-1.88}$. The highly skewed distribution of marriage distances is in this case well described by the model. However, as Boyce *et al.* (1967) note, this is not a direct test of the neighborhood knowledge model since they have no information on visiting patterns, the causal factor in their formulation.

Swedlund (1972) also found that the distribution of marriage distances in the 1810–19 population of Deerfield, Massachusetts, USA could be fit by the negative power function of the neighborhood knowledge model. The density of villages was lower in Massachusetts than in Oxfordshire, and Swedlund had to extend the geographic range of the neighborhood to 15 miles in order to include a comparable number of points ($n = 17$). The fitted curve for Deerfield was $y = 22.4x^{-1.05}$. The higher value of a

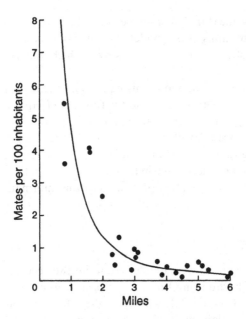

Figure 3.4. Relationship between mating frequency and distance in Oxfordshire, 1861. Circles represent villages contributing mates to the 1861 population of Charlton village. The expected distribution according to the model is the fitted curve $y = 4.75x^{-1.88}$. (From Boyce *et al.* 1967, fig. 1.)

in this population reflects the lower density of villages. The wide departure of the exponent from an expected value of two, however, suggests that differences other than merely density may be affecting this distribution. Again, no evidence on visiting was available for this historic population.

Since the model was proposed, only two studies have presented data on visiting frequencies in relation to mating distributions. These include my work on the Semai (Fix 1974) and Cavalli-Sforza & Hewlett (1982; Hewlett *et al.* 1982) on the Aka Pygmies. Neither of these populations is very similar to the densely settled agrarian society of rural England that provided the original impetus for the model. Both the Aka, central African hunter–gatherers, and the Semai, Malaysian swiddeners, have been described in Chapter 2. Both have densities very much lower than Oxfordshire or Massachusetts, and for the Aka, a fixed home base is probably lacking. Semai settlements are less fluid than for the Aka but still fission and fuse over a relatively short time span. Thus neither of these two societies were the best candidates to test the neighborhood knowledge model. Interest-

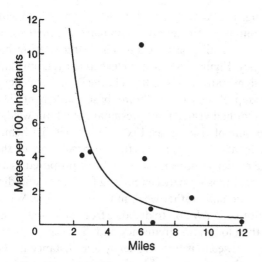

Figure 3.5. Relationship between mating frequency and distance in Semai study area. Circles represent settlements contributing mates to SA settlement. The fitted curve is $y = 34x^{-1.76}$. (From Fix 1974, fig. 1.)

ingly however, both provide evidence for the role of visiting and knowledge in determining mating range.

Case study – Semai

The low density of Semai settlements means that even within an extended range of 12 miles (*c.* 19 km) only eight exist. Referring to Figure 3.3 will also show that these settlements are not homogeneously distributed. Thus it should not be surprising that the Semai marriage distance data gave only equivocal support for the power function equation of the Boyce *et al.* model. Figure 3.5 shows a fitted line, $y = 34x^{-1.76}$, but a *t*-test of the slope of the log transformed data was non-significant ($p > 0.50$). The very few points are surely part of the reason for a lack of statistical significance but visual inspection of the data suggests other problems. Four settlements may be found at about 6.5 miles (*c.* 10.5 km) from SA, the 'home base' of the study. Notice that the mating frequencies among these four range from the highest value (*per capita*) to the lowest. The average of these values is not very different from that of the two settlements at three to four miles (*c.* 6 km) and only a little more than the settlement at nine miles (*c.* 14.5 km). There is a relationship between distance and mating frequency among the Semai (almost no mates are chosen from settlements more than 12 miles away), but within a neighborhood, that relationship is not exponentially declining. The reasons for this have not only to do with lower density, but also economic activities and socio-cultural practices.

Looking now at the Semai visiting pattern, a much clearer relationship exists with mating frequency. Figure 3.6 shows a clear linear relationship, where $y = 0.113x^{-1.39}$, significant at $p < 0.05$. The visiting frequencies were obtained from a group of men in the 'home base' settlement. The percentage of this group that had visited (and presumably had knowledge of) a settlement is the measure of visiting (see Fix 1974, for details). Semai men find mates in settlements they visit, exactly the prediction of the neighborhood knowledge model. However, the model was proposed as an explanation for the close relationship between mating frequency and distance, a relationship that is actually rather weak in the Semai case. When the visiting frequency is plotted in relation to distance from home-base, the pattern is even less clear than for mating frequency and distance.

The lack of correlation between visiting frequency and distance in the Semai case is due to the violation of several basic assumptions of movement patterns made by the model. The sparse population density and village distribution means that few intermediate villages are passed through on the way to visit more distant ones and people in the course of their normal activities would not be likely to travel through another village. The dense settlement of the English countryside and the economic activities of these farmers (for example, markets) are much more compatible with the Boyce *et al.* (1967) assumptions. Semai travel to other settlements consciously for particular purposes and usually to visit specific people. The reasons given for such trips include seeing relatives and attending ceremonies. Thus Semai usually are visiting localities where they have relatives, and the distribution of visiting mirrors the distribution of kin.

If the distribution of mating distances is to be explained as a result of familiarity gained by visiting, then the cause of visiting ought to be some factor other than prior familiarity or knowledge of a place. Economic activities such as marketing can initiate travel with the consequences suggested by the Boyce *et al.* model. (William Skinner 1964, for instance, has suggested that Chinese market areas are also marriage pools.) For the Semai, and I expect other kin-based societies, travelers already possess considerable knowledge about most places that they visit as they are visiting kin. Even if the kinsperson has not been met previously, the very fact of relatedness makes them 'known'. In this case, then, both mating and visiting may be caused by 'knowledge' of the wider neighborhood of kin. Thus a likely cause for a visit with consequences for future marriage might be the occasion of a curing ceremony held for a kinsperson in a neighboring

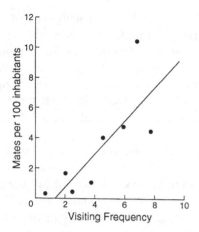

Figure 3.6. Relationship between mating frequency and visiting frequency in the Semai study area. Circles represent settlements. (From Fix 1974, fig. 2.)

settlement. Young men travel to 'help' the cure by singing and dancing, incidentally providing an opportunity to meet potential spouses.

The distribution of kin in the region is the result of prior history of settlement fission and fusion. Splinter groups found new settlements or fuse with established settlements (see Chapter 2 and previous sections for details). The salient observation for the neighborhood knowledge model is that these splinter groups may move some distance from the original parent village. For example, about 20 years before the study, a splinter group from a settlement some six miles (*c*. 10km) distant fused with SA settlement, the 'home base' in Figure 3.5. Individuals from the migrant generation and their SA-born offspring continue to visit this village regularly and intermarriage rates are the highest of all the settlements of the region irrespective of geographic distance. The neighborhood knowledge model was formulated to account for a particular distribution of mating distances typical of densely settled agrarian societies. It should not be surprising if the model can not be generalized to all populations. Nonetheless, the concept of information or knowledge as a determinant of human movement and dispersal is a powerful one (see Chapter 1) and clearly this is applicable to the Semai case.

Case study – Aka

Cavalli-Sforza & Hewlett (1982; Hewlett *et al.* 1982) defined a closely related concept, the 'exploration range', and applied it to the Aka Pygmies of the Central African Republic (see Chapter 2). They explicitly avoided the term 'neighborhood knowledge' as they felt that the specific model developed by Boyce *et al.* (1967) could not be applied to the Aka. The exploration range is defined as the distribution of probabilities of persons having visited localities at various distances from their home base. This range would certainly include the foraging range over which individuals hunt and gather but also includes a wider area that has been visited over the lifetime of individuals.

The Aka foragers are considerably more mobile than the Semai or rural English and their exploration ranges extend over much wider areas. Oxfordshire neighborhoods were limited to a radius of six miles (*c.* 10 km) and for the Semai, 12 miles (*c.* 19 km) encompassed the range of most visiting. The Aka, in contrast, had visited localities beyond 60 miles (*c.* 97 km) from their residences (Hewlett *et al.* 1982:421). Of relevance to mating considerations, most of Aka exploration seems to be accomplished by early adulthood providing opportunities for meeting potential spouses.

Also in contrast to the Semai, the distribution of visiting was well correlated with distance and could be fit with a negative exponential function of the form

$$p = e^{-x/k}$$

where p is the probability of having visited a place at distance x and k is a constant measuring mobility (Hewlett *et al.* 1982). The Aka are similar to the Semai, however, in usually traveling with a purpose in mind and, also as for the Semai, visiting kin was the chief reason given for travel.

The Aka case also confirms the relationship between knowledge of a region (gained through exploratory activity) and mating distance (Cavalli-Sforza & Hewlett 1982). The mean exploration range of the Aka males was about 45 miles (*c.* 72.5 km); the mean mating distance, 32 miles (*c.* 52 km), was less but similar. Since the distribution of exploration distances was an exponential rather than the power function of Boyce *et al.*, in the Aka case at least, the specific prediction of the neighborhood knowledge model is not met, although the actual difference between the two curves is not great.

Basic variables

The aim of this section is to identify several key factors that have not always been made explicit in the classic migration models. These are summarized in Table 3.2. Having now characterized the genetic models, we can examine these basic variables in the context of both the models and the case studies of human migration patterns. Particularly important are the implications for violating assumptions about these variables.

Life cycle timing

A variable rarely explicitly discussed is the timing of migration. When in the life cycle does migration occur, during childhood (pre-marital), at the time of marriage as one (or both) spouse moves and establishes a new conjugal family (marital), and/or following marriage during or after the child-bearing period (post-marital migration)?

In animals, most gene flow results from dispersal of young individuals (Merrell 1981). In plants, dispersal is generally at the seed stage by wind or animals. Larval organisms may spread vast distances as planktonic forms, mammalian young may be ejected from the parental range and seek their fortune elsewhere, and so forth.

Human dispersal can occur at any stage of the life cycle. Pre-marital movement is generally in the company of parents or other family rather than the individual-scattering characteristic of young plants or animals. Post-marital movement as a result of exogamy is enshrined in anthropological lore and forms part of the terminology all introductory students must learn (patri- or matrilocal, etc.). It is often this stage of the life cycle that is implicit in anthropological genetic models. On the other hand, many social science studies of migration focus on adults moving for economic reasons. These differences may not have major implications depending on the problem at issue. However, genetic models that presume dispersal occurs prior to population regulation (the large scale mortality characteristic of the very young) give different predictions than models deriving migrants from the survivors (Rogers & Harpending 1986).

In the cases discussed in Chapter 2, a wide range of stages of the life cycle were represented. In relatively sedentary societies such as the Gainj of New Guinea, most movement was a consequence of marriage, and women were the primary movers. This type of migration, with a clear referent to a specific event, is perhaps the most easily documented of any of the stages since data on spousal birthplaces are sufficient. In a classic system of patrilocal residence, where wives move to the homes of their husbands and no further movement takes place, parent–offspring birthplace data will also provide a complete description of the pattern. In fission–fusion societies such as the Yanomamo and Semai, significant movement may occur after marriage as groups of families move to new

Table 3.2. *Key assumptions in classic migration models*

Variable	Assumption	Implication of violation
(1) Life cycle	Gametic or infant stage	m a random variable with variance; migration stochastic
(2) Units	Gametes or individuals	Group migration may be structured
(3) Structure	Representative of donor population	Kin-structure of groups or chain migrants bias migrant gene frequencies
(4) Spatial pattern	Isotropic, 'frictionless plain'	Geographic barriers and patterned flows → spatial structure of genetic variation
(5) Distance	Distance primary (only) determinant. 'Long distance' migration stabilizing	Information and/or other factors affecting movement. Random long distance migrants
(6) Population sizes	Equal sizes for donor-recipient populations. Infinite pool for long distance migrants	Fluctuating populations → nonequilibrium. Expanding populations → colonization and/or demic diffusion. Stochastic long range migration

locations. Thus, for the Semai, one child might be born in the mother's natal village (or the current residence of mother's reference group kin), a second in the settlement of the father, a third in a new settlement following a population split, and so forth. This type of migration necessarily involves pre-marital displacement for the young children moving with their parents. Similarly, the job-seeking mobility of industrial societies and the greater mobility of tenants in societies such as the Basque carry along children to the parents' destinations willy-nilly. The very high mobility of some hunter–gatherers (e.g., !Kung) also involves all the life cycle stages. Much of this movement, however, might better be considered ranging or foraging of members of a dispersed population rather than genetic displacement.

Classical genetics models ignore these differences in the timing of migration. Time is measured in generational units in these models and the relevant displacements are between the population of birth and the population within which reproduction takes place (most easily represented by parent–offspring birthplace differences as in the migration matrix approach). A great deal of the mobility of individuals is simply irrelevant to population genetics if it has no effect on the intergenerational movement of genes. A world traveler who returns to marry and reproduce in his natal village is not a migrant in the genetic sense (even though such wide geographic knowledge might be expected to expand his potential spouse pool and make his local marriage less likely). However, it has recently been pointed out that genetic models that presume that dispersal occurs prior to population regulation (the large scale mortality characteristic of the very young) give different predictions than models deriving migrants from the survivors (Rogers 1988; Rogers & Harpending 1986). In particular, the amount of genetic variation expected among subdivisions of a population in a migration matrix model will be affected.

Rogers & Harpending (1986) contrast two models of dispersal. Model A corresponds to the life cycle characteristic of most plants and many animals where dispersal of pollen or seeds or juveniles occurs before the high mortality of this life stage has taken place. Model B is more appropriate for humans (at least for those societies where most migration is marital) in presuming that migration does not occur until after the high mortality phase of life.

Model A

	migration		regulation		reproduction	
Newborns	$\rightarrow$	Adults	$\rightarrow$	Adults	$\rightarrow$	Newborns
(∞)		(∞)		(n)		(∞)

Model B

	regulation		migration		reproduction	
Zygotes	$\rightarrow$	Newborns	$\rightarrow$	Adults	$\rightarrow$	Zygotes
(∞)		(∞)		(n)		(∞)

where in both models the population size is in parentheses under the life stage and the processes of migration, population regulation, and reproduction are represented above the arrows.

Following Wright (1931), population genetics models of migration assume the life cycle pattern of Model A. An essentially infinite number of 'newborns' (gametes, seeds, juveniles) disperse and then experience high mortality as most fail to establish themselves in new habitats. The adult population (now of size n) reproduces and the cycle continues. Genetic drift in this model occurs as a consequence of the reduction of the population size *after* migration has occurred. Since this model presumes a very large number of migrants, the gene frequency of migrants has no variance and therefore migration does not contribute to genetic variation among subdivisions but only reduces such variation.

Model B assumes that migration takes place only after a stage of high infant mortality that reduces the population size to its adult size (n). This pattern fits human life history much better than Model A. Although some human infants and juveniles migrate along with their families, the primary stage of movement is young adulthood long after the period of high infant mortality. Genetic drift has already occurred prior to migration as a result of gametic sampling and random mortality of infants (note also that natural selection may be occurring during this life stage as susceptible infants succumb to malaria and other diseases). The critical difference between the models is that migration in Model B occurs after the population has been reduced to a finite (perhaps small) size, therefore, a variance component is added by migration since migrants are now samples of a population.

Rogers & Harpending (1986) go on to develop a revised migration matrix approach based on a Model B life cycle migration pattern. They note that previous theoretical work had considered the effects of random variation among migrants under the rubric of 'stochastic migration' (Nagylaki 1979; Sved & Latter 1977) but that these models were based on the unrealistic assumptions of the island model. (It might be noted, however, that computer simulation models to be discussed below (e.g., Fix, 1978) implicitly followed a Model B approach, in that the order of operations put migration after genetic drift.) The major conclusion of Rogers and Harpending's model is that the genetic variance among newborns will be substantially greater (up to twice as much depending on the migration rate) than for adults. Thus genetic sampling of all age classes of the population commonly practiced in studies of human variation will inflate estimates of genetic variation among subdivisions compared to those made on adult gene frequencies alone.

Units

The classic genetics models reduce migration to the proportion of genes introduced by migrants to a population irrespective of whether the migrating units are gametes, individual organisms, or groups of individuals. Just as the implicit presumption for the timing of migration at the gametic or early life cycle stage, the unstated assumption was that individual gametes or organisms are the dispersers. However, as we have seen in the case studies, the total amount of migration between two localities may comprise the sum of movements by individuals at marriage or for economic reasons and/or it may encompass families or groups of families moving together following a population fission as for the Yanomamo and the Semai.

The relevance of the units of migration to genetic models is that migrant groups are less likely to be a random or representative sample of the donor gene pool from which they came. Individual migrants might represent such a random sample; even so, when numbers are small, even a random sample may deviate from the population mean. Thus a sample of 10 random migrant individuals from a population might very well not have the same frequency of alleles at all loci as their natal gene pool. Migration is always stochastic in this sense. The probability of deviation from the population mean will be a function of the number of migrants. As for all such sampling events, small numbers of migrants would be more likely to fluctuate from the mean.

Consider a commonly used formula representing gene frequency change due to migration in a stepping stone model:

$$q'_i = (1 - m)q_i + 0.5mq_{i-1} + 0.5mq_{i+1}$$

where q'_i is the gene frequency in the ith population after migration and m is the migration rate apportioned as one-half from each adjacent population. This formulation assumes that the contribution of migrants to the gene frequency in the following generation is simply a proportion (depending on their numbers) of the gene frequency in the population from which they are derived. Under these circumstances, migration will always act as a deterministic force moving gene frequencies in the recipient population closer to those in the donor population. But, clearly, this effect depends on the migrant gene frequency being equal to that of the parental population. The smaller the scale of migration, the less likely is this condition to be true. Certainly Wright (1955) was aware of this possibility, listing 'fluctuations in immigration effect' as one of the 'random processes' in his classification of the factors of evolution. More recently, others (Nagylaki 1979; Rogers 1988) have modeled migration as a stochastic force.

Kin structure

The classic models also fail to consider the potential effects of the scale of migration for the structure of migration. When the units of migration are groups, they may be kin-structured. That is, rather than migration representing a random sample of individuals or gametes from the population, groups of biologically related kin may be a highly biased sample of the natal gene pool. The likelihood of significant deviations from parental gene frequencies in kin-structured migrant groups is a consequence of the shared genetic ancestry of biological relatives. Depending on the degree of relatedness, kin are more likely to be genetically similar than random individuals. Thus, full sibs share on average half their genes by inheritance from their two parents; half-sibs share only one parent and therefore a quarter of their genes should be identical by descent; first cousins, one-eighth, and so on.

Viewed as a sampling phenomenon, a group of relatives is a potentially biased sample since the individuals making up the group are not independent units. A group comprised entirely of sibs can only represent their parents' genes, a total of four genes for any locus rather than the $2N$ genes present in the population as a whole. Analogously to genetic drift, another sampling process, the smaller the population, the greater the potential deviation. Since kin-structuring reduces the number of independent genomes in the sample of individuals making up the migrant gene pool, it increases the variance of the migrant gene frequency. For a one-locus, two-allele system with gene frequencies p and q, binomial variance of sampling for the gene frequency of the migrant group will be $\sigma^2 = pq/2N$, where N is the group size. However, where the individuals in the group are genetically related, the number of independent genes is not $2N$, twice the group size, but a reduced number dependent on the degree of relatedness among the group. This consideration emphasizes the difference between measuring the number of migrants being exchanged among subdivisions and the *genetically independent* number of migrants. Depending on the degree of kin-structuring, the latter number may be much smaller than simply the total of individuals migrating.

The consequence of kin-structured migration (KSM) is the augmentation of the stochasticity of migration. Migration modeled as a deterministic process such that migrant gene frequencies are identical to the population-of-origin gene frequencies will always reduce the variation among subdivisions. Migration modeled as a stochastic process (Epperson 1994; Nagylaki 1979; Sved & Latter 1977) where migrant gene frequencies are random variables, may be less effective in reducing genetic differentiation. KSM may actually *increase* local genetic variation (Fix 1978). For rare alleles, all copies in a population may be confined to a single family group. Emigration of this family would remove the allele from the population.

Rogers (1987) has developed a method to evaluate the magnitude and effect of KSM. His model follows the migration matrix approach and, as for the parent model, aims to predict the amount of genetic variation expected among a set of exchanging local groups. The original paper may be consulted for the details of the full model, but the primary interest in the present context is the treatment of KSM.

The equation (equation 5 in Rogers 1987) describing the effect of KSM on genetic covariances among local groups is

$$\frac{\rho}{1 - \rho} \cong \frac{1}{4\bar{n}m_e} + \frac{\theta}{2\bar{n}}$$

where ρ is the expectation of the normalized covariance of gene frequencies in the local groups and is equivalent to Wright's (1951) F_{ST}, $\bar{n}$ is the average population size of the groups, m_e is the effective migration rate among groups, and θ, the parameter of particular interest here, is a measure of the degree of kin-structuring of migrants.

Rogers shows that the value of θ is a function of the size of related migrant groups and the genetic correlation among individuals within the groups. Thus if the size of the 'family' (= kin group whatever its structure) is γ, and the within family genetic correlation is κ, then $\theta = (\gamma - 1)\kappa$. This result can be extended to the case where family groups vary in size.

The effect of KSM on genetic variation depends on the value of θ in this formulation. This can be substantial. For instance, if pairs of sibs are the migrant groups, $\gamma = 2$, $\kappa = 0.5$ and $\theta = (2 - 1)0.05 = 0.5$. Rogers notes that when family group size varies, γ will be larger than the mean family size. A few large families can increase γ greatly. Rogers gives an example where 90 percent of families are two individuals and 10 percent include 15 persons; the mean family size of this assemblage is 3.3, whereas the value of γ is 7.9. Thus, variation in family size can inflate γ and thereby the effect of KSM. On the other hand, as group size increases, the correlation among individuals might be expected to be less (reduced κ) decreasing the effect of KSM.

The effect of KSM is conditioned by the migration rate as well. Thus the absolute increase in the amount of genetic variation among groups (compared to the expected variance within groups) is a function of both θ and m_e in Rogers' model. This leads to the conclusion that KSM should have greater impact in populations with high rates of migration. Taking all these factors into account, Rogers suggests that the maximum effect of KSM for humans might occur when entire sibships migrate together. Assuming family sizes average four persons, this form of migration could increase the ratio of among to within group variance by as much as three-fold.

Bryan Epperson (1993; 1994) has further considered the consequences of stochastic migration and KSM in the context of his general reformulation of migration-drift models as space–time autoregressive moving average processes. His conclusions are generally similar to previous studies on stochastic migration in emphasizing the increased variance in spatial correlations among adult subpopulations. In addition, he goes on to show that a critical factor determining the impact of stochastic migration is the degree to which the migration effects are *shared* among subpopulations.

Where each migrant sample is independent of every other such group, stochastic changes affect only the recipient population. This does serve to increase the overall variance of the system of subpopulations and migration of this sort can be categorized as another component of genetic drift (Rogers 1988). However, Epperson (1994) suggests that in many real populations, migrant groups originating in the same local group and moving to different nearby groups may not be genetically independent of one another. As a consequence, more than one adjacent subpopulation may receive *correlated* genetic inputs from a migrant source. Such shared stochastic migrant effects can dramatically change spatial patterns of genetic variation.

Figure 3.7 illustrates the potential magnitude of this effect based on a linear stepping stone model (nearest neighbor exchange only). Basic parameters for this case were: local migration of 0.0405; and systemic force (m_∞) of 0.01. The vertical axis is the genetic correlation at the distance class represented on the horizontal axis. The middle curve (connecting triangles) depicts the usual expectation for a smooth decline of genetic correlation with distance under a genetic drift-deterministic migration stepping stone model. Migrant gene frequencies are identical to donor population gene frequencies following the normal assumption in such models and migration increases genetic correlation between exchanging populations. The upper curve represents positively shared stochastic migrant inputs, a process that markedly increases the correlations between near neighbors since they are receiving correlated gene inputs. The lower curve shows the dramatic effect of *negatively* shared migration. Under this model, gene frequencies of migrant groups from the parental subpopulation are more dissimilar than random groups. Potentially, one adjacent neighboring population could receive a migrant group with a *higher* allele frequency than that in the donor group while the other neighbor could receive a migrant sample with a much *lower* gene frequency. This situation completely contravenes the conventional expectation for the role of migration and leads to a negative correlation between distance and genetic correlation.

These shared stochastic effects might arise in a variety of ways. Positive sharing could result from an initial stochastic sampling producing a pool of emigrants from which are formed the migrant groups that go to different subpopulations (obviously, this division could be the cause of a further stochastic sampling event). Negative sharing occurs when migrant groups show negative conditional genetic correlations. Epperson (1994:185) suggests that this is exactly the kind of situation represented by fission–fusion populations such as the Yanomamo and the Semai.

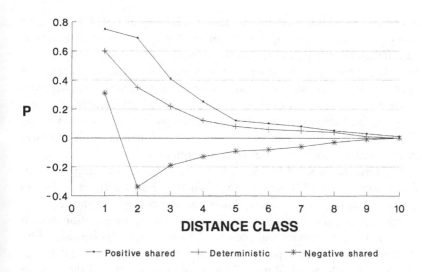

Figure 3.7. Comparison of spatial correlations (P) for three different patterns of migration: (1) line connecting squares are models with positively shared stochastic migration; (2) line connecting pluses are models with unshared stochastic inputs; (3) line connecting asterisks representing models with negatively shared stochastic migration (fission) inputs. (Redrawn from Epperson 1994, fig. 3.)

Kin-structuring of migrant groups is particularly important in this formulation. Epperson (1994:172) shows that when migrant groups are formed as random samples of the donor gene pool, the shared effects of stochastic migration are less salient but as kin-structuring becomes more pronounced, shared effects 'can dominate the form of the spatial and space–time correlations' (Epperson 1994:173).

Kin-structured migrant groups may occur in either positive or negatively shared contexts. Taking the example of the Semai, the fission of a settlement due to some disaster such as an epidemic might divide people into kin groups (likely to be more dissimilar to each other than a random division of the population) which then fuse with different neighboring settlements, an example of negative sharing. Equally, a kin group (more similar in their genetic constitution) might decide to emigrate as a result of a dispute, some members of which going to one neighboring group, the others to a different settlement, illustrating positive shared migrant inputs.

Clearly, Epperson's results have important implications for our understanding of spatial variation in gene frequencies. Unshared stochastic migration effects can increase the genetic variance among subpopulations substantially, if emigrant sampling is nonrandom or kin-structured. In contrast to expectations that migration will homogenize allele frequencies among exchanging subpopulations, increasing rates of stochastic migration actually *increase* variation. Perhaps even more striking is the observation that shared negative stochastic migration may produce a pattern of spatial genetic correlations exactly opposite to that of the classic stepping-stone or IBD models. Rather than genetic correlations decreasing monotonically with distance, spatial correlation may increase after an initial sharp decline at close distances. (Some of the consequences of this effect will be explored by computer simulation in chapter 4.)

The *potential* of KSM to affect genetic variation substantially has been demonstrated (Fix 1978; Rogers 1987). Whether this potential is realized in actual populations is an empirical question and depends on the degree to which real migrant groups are kin-structured and the distribution of such migration among species and societies. The two examples considered by Rogers (1987) – lions and pine voles – differed in the magnitude of the effect of KSM. The dispersal of male lions in sib groups inflated the ratio of among to within group variation by a factor of 1.4, while pine voles showed only slight effects of kinship. Consideration of the dispersal mechanisms of many plants suggests that some species may show extreme KSM effects (Levin & Fix 1989). The great heterogeneity among human societies in size and structure can be expected to produce corresponding variation in the role of KSM.

As Rogers' (1987) analysis makes clear, KSM should be more important in groups with high migration rates. Human societies vary greatly in mobility, as the case studies of Chapter 2 demonstrated. Although the role of kinship varies in different societies, it has an important role to play in each. In larger polities, other principles become important. Nonetheless, as the discussion in Chapter 1 and many examples in Chapter 2 illustrated, family considerations are often central in migration.

Two of the societies described in Chapter 2, the Yanomamo and the Semai, have been prime exemplars of KSM. Both are small scale societies in which kinship is a major organizing principle and in both, group migration is frequent. The politics of village fission in the Yanomamo case produces kin-structured groups that may either fuse with another village (migration) or found a new village. The latter event was emphasized by Neel and his colleagues (e.g. Neel & Salzano, 1967) and labeled *lineal effect* to emphasize the role of the unilineal kinship systems of South American Indian societies such as the Yanomamo. This process may augment the stochastic founder effect in the same way that KSM increases the stochasticity of migration. That unilineal kinship systems are not crucial to the effect is shown by the Semai. The fission–fusion social structure described for the South American Indian tribes splits along patrilineage lines. Semai settlements, though lacking unilineal kin groups, split, and family groups, often sib groups but sometimes with complex bilateral ties, migrate to new areas (founding new settlements) or fuse with extant villages. The basic process is very similar to the Yanomamo case except that they do not split along patrilineages. KSM depends only on the fact that biological kin are likely to share many genes in common. Therefore, *any* group of families, whatever their kinship reckoning system, will be more similar genetically than a random group of individuals. Thus 'lineal effect' is not limited to societies with unilineal kinship systems.

Various degrees of KSM have been documented in populations spanning the variation seen in Chapter 2 in densities and modes of social integration. For example, Roberts (1968) recognized kinship factors in emigration from Tristan da Cunha, a population of English descent and bilateral kin reckoning. McCullough & Barton (1991) examined relatedness among the colonists in the founding populations of Plymouth from 1620 to 1623. They found strong evidence for co-migration of nuclear families but that links among such families were not common and many individuals (especially males) were unrelated to any other colonists. This lack of strong KSM may be related to the mode of recruitment of the colonists. Rather than representing a single residential community, the colonists were broadly drawn from several regions of England (McCullough & Barton 1991:362) – from Yorkshire to Devon. Kinship (except for nuclear family ties) thus was a less important integrating force in this society than their common religion.

A strong religious ideology also structures the migration of several of the well-known 'Old Order' brethren of the Americas, although kinship does seem to play a role in group migration among some of these groups. Schisms resulting in the formation of new colonies among the Amish are influenced by family (Hurd 1983). The Amish situation contrasts with the Plymouth case as colony fission occurred after extended periods of coresidence and the development of family ties. As a consequence, the mean coefficient of relationship calculated by Hurd for the Amish (0.065) is not very different from the equivalent measure on the Yanomamo (average 0.086; Chagnon 1988). Sufficient kin are available to form kin-structured groups, but for the Plymouth colonists, little previous kinship existed among persons widely separated in place of origin.

The Hutterites, a communal religious group similar to the Amish, also periodically fission and establish new colonies. From an initial small founding population arriving in the United States in 1874, rapid population growth and colony fission has produced some 309 daughter colonies distributed through the Plains of Canada and the United States (Olsen 1987). Olsen explicitly compares Hutterite fission to the Yanomamo case. She points out that despite the enormous cultural differences that exist between these two populations, both the Hutterites and the Yanomamo are growing populations that expand into new territory by fissioning and colonizing. Average new village (colony) sizes and maximum population sizes at the time of fissioning, 75 and 250 individuals respectively, are similar in the Yanomamo and the Hutterites. Also similar to the Yanomamo, Hutterite fission groups are kin-structured, although not by patrilineages. Mange (1964), for example, described one founding group of 68 persons that included 5 sibs, 33 half-sibs or uncle/nephew, and 24 first cousin relationships.

Migration that is non-random with respect to kinship is not limited to positive associations of migrating kin. Leslie (1980) has shown that on the island of St. Barthelemy, French West Indies, individuals who marry exogamously from their natal into one of the other parishes are more likely to have close kin ties to their local potential mates than are those who marry endogamously. Migrants are therefore *less* closely related to each other than would be a random sample of individuals drawn from the natal gene pool. As Leslie points out, the genetic consequences of this negatively KSM would be a reduction of inbreeding in the local community and increased homogenization among the parishes of the island.

Despite the demonstration of genealogical ties among migrants in many cultures, KSM is unlikely to be universal. Thus when Manderschied et al. (1994) ask the question, 'Is migration kin structured?', the answer is, 'Yes and no'. Low mobility populations in which groups rarely migrate together or, conversely, high mobility groups drawing individual migrants from a variety of areas, would not be expected to show significant levels of KSM. Expanding populations such as the Yanomamo and Hutterites and fission–fusion societies such as the Semai clearly do. As for most variables, the degree of KSM should vary.

A further issue, however, arises – that is, is KSM demonstrable using a method other than genealogy? Rogers & Jorde (1987) devised a method to infer KSM from genetic data and it is this method that is being employed in the Manderscheid et al. (1994) article. This procedure requires that statistically significant genetic differences must exist in the estimator for kin-structure. Statistical significance depends on sampling sizes and other considerations. Further, gene frequencies are the outcomes of the interaction of many potential causes, including natural selection, mutation, gene flow, and genetic drift. As the long controversy in population genetics over the relative roles of natural selection and genetic drift demonstrated, inferring process from gene frequency distributions is an uncertain procedure (Lewontin 1974).

The estimator of kin structure, α_M, derived in Rogers & Eriksson (Rogers 1988; Rogers & Eriksson 1988), is approximately equivalent to the previously defined measure, θ (Rogers 1987, 1988: Rogers & Eriksson 1988). A new term, *migrule*, is used in Rogers & Eriksson (1988) to refer to the migrant groups called families in the 1987 paper. α_M is a measure of the degree to which the variance in allele frequencies in kin-structured migrules is increased over the corresponding variance in a random group of migrants.

The estimation of this quantity requires data for each individual on the birthplace, adult residence, and phenotypes for as many loci as possible. The individuals who were born in the same place, i, and who now reside in a new place, j, as adults are considered members of the ijth migrant set of size N_{ij}. If these groups are kin-structured these individuals should be more similar genetically than would be a random group of migrants from place i. The estimate of α_M is obtained by comparing the sums of squares of allele frequencies in migrant groups with those of all the individuals in the birthplace from which the migrant group originated (see Rogers, 1988:451, for formulae).

The composite estimator is obtained from information on all birthplaces and all alleles. It seems reasonable to assume that kin structuring should affect all loci equally, and that there should be no particular reason that kin structuring should vary between places, therefore each locus and each place should be providing an estimate of the same quantity. The sampling distribution of α_M can be approximated by a Monte Carlo simulation using the actual data on individuals in the sample. One thousand repetitions of the simulation generates the distribution under the null hypothesis that $\alpha_M = 0$.

Rogers & Eriksson (1988) apply their method to data from the Åland Islands, Finland. This population has been extensively studied (summarized in Jorde *et al.* 1982; Mielke *et al.* 1976) and extensive data on phenotypes as well as birthplaces and residences are available. Recalling that KSM is most likely in populations with high mobility rates, the Åland case seems unpropitious for demonstrating its effect since migration among the subdivisions of the islands is not common. In fact, nearly all the migrant 'groups' comprise from zero to five persons. Not surprisingly then, Rogers and Eriksson fail to find a significant effect of kin structure.

Manderscheid *et al.* (1994) extend this application to five more human populations and one rhesus macaque population. None of these populations yielded estimates of α_M that were significantly different from 0 leading them to conclude that these results provide no evidence for KSM being a widespread phenomenon. However, there is a difference between 'providing no evidence for' and 'disproving'. As the authors note, there are several reasons why their test may underestimate the pervasiveness of KSM.

One problem is that the statistical power of the test is not great. A value of α_M would not be deemed significant with these data unless its true value was approximately 0.4. Recall that if migration always comprised a pair of sibs, α_M would equal 0.5. One of their human populations is shown with an estimated α_M of nearly 0.4 suggesting the possibility of at least mild kin-structuring in this case; however, this value is still not significant.

Further, they point out that there are several potential sources of bias in these data, two of which result from the sampling of the population as genetic data are collected. A classic phrase to describe genetic samples is 'apparently unrelated individuals'. Clearly this sampling scheme would fail to include close relatives that may have migrated together thereby reducing the estimated value of α_M. The other side of the coin would be the situation where too many relatives are included in the genetic sample. Such a procedure would bias the expectation for random migrants again reducing the apparent estimate for α_M. These potential sources of bias are not unlikely given the 'fairly informal approach to sampling' (Manderscheid *et al.* 1994:56) adopted by many field investigators. The reality of small populations is that relatives often cluster in their residences and/or their likelihood of coming forward to donate blood, therefore, kin-structured genetic samples may be fairly common.

The most likely explanation for the low values of α_M found by Manderscheid *et al.* is simply that these populations represent a very small sample of the human range and they are all highly localized to one culture area – five of the six are from Papua New Guinea. The very low rate of migration in the Åland case has already been noted. One of the Papua New Guinea societies in the sample, the Gainj, was described in Chapter 2. It will be remembered that the Gainj were very highly localized with mean father–offspring birthplace distances of only 1.3 km. Almost all movement was of women at the time of their marriage moving to their husbands' villages. Density was high and movement of groups of kin either to new locations or to fuse with established villages would be very difficult if not impossible. These are clearly not the circumstances under which KSM would be predicted. Since the other New Guinea groups can be expected to share many of these basic characteristics, they too would be unlikely to show KSM.

KSM refers to the continuing exchange of migrants between established local groups. The lineal effect, originally proposed by Neel & Salzano (1967), can be seen as simply kin-structured founder effect – that is, an augmented founder effect as new villages are settled by kin-structured splinter groups. Since the gene frequencies in the new group are a biased sample of the donor gene pool, they may diverge markedly. Neel & Salzano (1967) stated that this could be an important cause for the high level of genetic microdifferentiation observed in the South American groups. Fix & Lie-Injo (1975) showed that even alleles likely to be subject to strong natural selection (e.g. hemoglobin E, ovalocytosis) varied considerably among Semai settlements, a finding consistent with the fission–fusion structure of Semai society (Fix 1975).

Smouse *et al.* (1981) studied the effect of lineal fission compared to random splits on genetic divergence in the Yanomamo. They found that the initial differences between two villages after a lineal founding event were very large. They were able to show that the impact of the non-random kin-based structure of the splinter groups was profound, reducing the effective size of villages at the time of fission by a factor of four, relative to expectation from random division.

The consequences of the lineal effect, beyond the founding gene frequency differences, depend on the subsequent history of migration and gene flow between the new population and its parent and neighbor populations. If exchange rates are high, initial difference might be rapidly swamped by migrants. For instance, one of the Yanomamo splits observed by Smouse *et al.* (1981) involved a 'friendly fission' where the usual political strife causing the departure of a dissident group was less pronounced and continuing marital exchange occurred after division. In this case, the gene frequencies in the new village were hardly more divergent than expected from random fission. Rogers & Harpending (1986; see also Rogers 1987) pointed out that in populations where fissions are infrequent and convergence toward drift/migration equilibrium is rapid, the effects of lineal fission will be transitory. On the other hand, Neel & Salzano (1967) speculated that lineal effect might have played a crucial role in the genetic differentiation of new tribes as migrant kin groups became isolated. Particularly as populations expanded and extended their ranges, opportunities for lineal effect (or kin-structured founder effect) could have arisen (see Chapter 4 for a simulation model of this phenomenon).

It should also be noted that because kin-structured founder effect is a stochastic process, the gene frequencies of newly founded groups would be likely to differ from those of the parental populations, compounding the difficulty of inferring population trees of descent from contemporary genetic patterns.

Both lineal effect and KSM focused on the genetic effects in the recipient (or newly founded) rather than the donor population. However, as Roberts

(1968) pointed out, the *emigration* of families from a small population can have a great effect on gene frequencies of the remaining population. The departure of random individuals from a tiny isolate such as Tristan da Cunha exerts a stochastic change in gene frequencies. Roberts showed that in the history of Tristan, most often it was family groups that departed together, a phenomenon he labeled *booster effect*, in recognition of increased chance of deviation in gene frequencies due to kin structure. In general, kin-structuring of migrant groups, especially when these groups are derived from small populations, may influence gene frequencies in the parental group (by removing a non-random sample of alleles), in the recipient population, and/or a newly founded population.

While group migration may increase the likelihood of kin-structuring, a series of individual migration events over time may still show kin-biasing. The sequential migration of families has been extensively documented in the literature on migration (De Jong & Gardner 1981) and has been labeled *chain migration*. The opening of a channel of migration may be instigated by one family member who is followed by others as conditions allow. Viewed in generational time, such chain migration of kin would be no different in its genetic effect than the movement of a kin-structured group.

Spatial pattern

The pattern of migration in space is a variable often explicitly recognized in migration models. Classic genetics models broadly encompass the different patterns of migration from continuous distributions of population and isolation by distance to stepping stone and island models of population structure. The dimensionality of migration is usually incorporated in genetic models as well, linear migration being contrasted with migration in two dimensions. Mathematically, the decline of genetic correlation with distance in models such as the stepping-stone is strongly affected by the number of dimensions (see figure 9.92 in Crow & Kimura, 1970).

Geography and distance

Population genetics models share with most geographic models of migration the commitment to distance as the primary (in some models, the only) determinant of the amount of movement between populations.

Only the island model ignores the central importance of distance in determining rates of migration between populations. This focus is consistent with the general models of migration devised by geographers (see

Chapter 1); the invariant geographic variable in the study of migration is map distance.

At the same time, other geographic barriers can be expected to affect migration. Part of the isolation characteristic of some island populations is due to simple distance from other populations. Oceans may, however, serve as impediments to travel and reduce movement or, where suitable technology exists as in Oceania, allow movement over great distances. Similarly, mountain ranges may limit migration. While the classic population genetics models ignore these barriers, computer simulation techniques discussed in Chapter 4 permit greater geographic reality to be modeled. Barbujani *et al.* (1995), for example, incorporated several major geographic features (mountain ranges, seas) in their simulation of European populations.

A common feature of most genetic models including the island, Malécot's IBD, and migration matrix, is a stabilizing force to counteract genetic drift in the subdivisions and prevent fixation. This force is usually labeled 'long distance migration' and is often operationalized as migration from outside the study area (recall the discussion of this parameter along with the Malécot IBD model above). These long distance migrants are assumed to be members of a 'continental' population which is panmictic and large enough to avoid genetic drift of gene frequencies. Conveniently, the gene frequencies on the continent are simply the average of gene frequencies in the local populations under study. These continental gene frequencies are assumed to remain constant through time thereby counteracting the tendency of small subdivisions to drift away from the overall mean gene frequencies of the total assemblage.

Perhaps nothing contributes more to the surreal quality of some of these models than this assumption. When some attempt is made to actually study 'outside' migrants, such as the analysis of Makiritare Indian data by Wagener (1973), the 'continent' turns out to be simply other small populations located at greater distances but otherwise indistinguishable from the local 'subdivisions'. In the Makiritare case, outside groups included other Indian villages from neighboring tribes (including the Yanomamo). 'Outside' migration is merely 'unstudied' migration and gene frequencies in these groups can be expected to behave in the same way as the local subdivisions. To treat these migrants as representatives of an infinite panmictic population with invariant gene frequencies seems hard to justify.

While in principle this long distance migration (or some systematic force) is required for equilibrium, most empirical studies of human migration have focused on local movements. As Harpending & Ward (1982) note, where short-term patterning is the topic of interest and migration from outside is not high, this stabilizing force is not critical to understanding genetic variation among subdivisions.

At the same time, the history of human populations is filled with large scale movements and long range trade surely extends far back in time. Distant populations for a variety of evolutionary reasons are likely to differ genetically from each other. Thus the pattern of movement may have strong genetic implications. For instance, Livingstone's (1989) computer simulations (discussed in Chapter 4) demonstrate the important effects of long distance migrants in increasing the rate of spread of the hemoglobin alleles.

Population sizes

Most genetic models assume equal exchanges of migrants among equal sized groups. Further, constancy of population sizes and migration rates is assumed to persist through time. The classic discrete population models, such as the stepping stone, assume constant population sizes and equal exchange over the duration of time necessary to achieve genetic equilibrium. The migration matrix approach does not depend on equal exchange, using instead the actual rates, found empirically, of migration among colonies. However, these rates must be balanced over the long term, otherwise, populations will build up in one region at the expense of others, a point made by Wood (1977).

As we have seen in the various cases considered in Chapter 2 and the examples illustrating the application of the genetic models, the influences on human migration patterns rarely remain constant over even short periods of time. Certainly in recent times, the rise of cities and industrial growth have profoundly affected population sizes and migratory flows. However, even the very stable horticultural populations of Bougainville Island studied by Friedlaender (1975) have not maintained constancy of migration pattern and rates for the scores of generations required in some genetic models. For societies such as the Yanomamo and Semai, even over the short-term of one or two generations, fission and fusion will occur rearranging populations in size and often their spatial location. Thus in real populations through history relative population sizes and number of migrants contributed by donor and recipient populations must have varied along a continuum from the uniformity and constancy assumed in stepping stone models to various imbalances in population size and migration rate to the case where a donor population sends colonists into unoccupied territory.

When circumstances lead to sustained population growth by some populations, range expansion involving the movement of colonists into unoccupied territory may occur. This 'migratory' process presents the opportunity for founder effects perhaps augmented by lineal effects (Neel & Salzano 1967). Where technological or other advantages allow differential population growth at the expense of neighbors, expansion (or 'invasion' in Weiss', 1988, terms) into territory already occupied by others is possible. Cavalli-Sforza and his colleagues (1993) have dubbed this process 'demic diffusion' and have attributed major genetic consequences to it. (Their claim will be evaluated in Chapter 5.)

In the next chapter, we shall continue to explore the consequences of modifying key assumptions of the classic population genetics models and the roles of variables such as kin-structure using more complex computer simulation models.

4   *Computer simulation models*

Experimental approaches to history and the interaction of evolutionary forces

A recurrent theme of the previous chapter concerned the trade-off between simplifying complex events in order to model them mathematically and thereby ignoring potentially crucial determinants of the process. Thus the classic algebraic models of evolution reduce the interaction of genetic drift and migration to two parameters, N_e, the effective population size, and m, the migration rate. However, as the discussion at the end of the last chapter indicated, many factors not considered in these models may influence genetic variation and the course of microevolution in human populations. Adding variables to analytic models may encompass some of these factors as, for example, Rogers (1987) was able to incorporate kin-structure into his mathematical model of migration. Too many such additions result in the loss of mathematical simplicity and clarity.

Computer simulation provides an alternative method to explore more complex interactions among processes. Models may be made as complicated as the investigator desires. Indeed, there is often a great temptation to needlessly add procedures and variables to a simulation model because it is so easy. The greater the complexity of the model, the more difficult it may be to interpret the outcomes as too many variables may influence the results. Thus simulation modeling allows greater realism in the study of complex phenomena at the cost of loss of generality and the potential cost of inability to discriminate among a plethora of causal factors (Dyke 1981).

Bennett Dyke (1981), in his review of computer simulation in anthropology, has provided guidelines for the appropriate use of the method. In his view the most useful application of simulation is to generate artificial 'data'. Often the actual observational data collected in anthropological contexts are incomplete or pertain to only a short time period. Much more extensive simulated data produced under controlled conditions allow comparison with analytic models of the same process as well as the evaluation of statistical properties of the data distribution.

94

Although the idea of *artificial data* may seem counterintuitive (or at least counterempirical), all experimental results could equally be labeled 'artificial' in that they are produced by manipulation outside the 'natural' setting under controlled conditions. Simulation is a kind of experimental method with the computer substituting for the lab bench.

The data generated by any simulation program depends on the numerical values of the input parameters. Models incorporating many processes and variables require numerous replicate experiments using different parameter values. Clearly not *all* combinations of parameter values need be tested but 'exploring parameter space' in a series of replications by bracketing the high and low ends of likely variable values is important.

Simulation is particularly appropriate to the study of the population structure of small-scale human groups since it allows exploration of the stochastic variation in processes to be expected in small populations. The classic algebraic models of Chapter 3 provide predictions for steady state distributions of genetic variation, after some unspecified passage of time when genetic drift and migration reach equilibrium. However, in small populations, random fluctuations may have significant effects on the rate of approach to equilibrium and/or the final outcome. Simulation of random processes further emphasizes the need for multiple repeats of computer runs in order to chart the distributions of output data.

These general properties of simulation models will be apparent in the sections which follow. Several examples of studies of migration and microevolution ranging from the minimum size of endogamous populations through various models incorporating interactions of migration and the other evolutionary forces will be surveyed.

Minimum endogamous population size

Mendelian populations are defined as breeding populations and endogamy or mating within the group is the fundamental property of such populations. For species, endogamy is absolute; the criterion of the biological species is reproductive isolation from all other populations. As species habitats have become more fragmented and their populations reduced through human activities, conservationists have become more concerned about species' minimum viable population sizes (Soulé 1987). Beyond a critical lower population limit, lack of mates might mean extinction for the species.

Within a single species, local populations may be *relatively* rather than absolutely endogamous so that a majority of mates are found within the local group. Clearly the consequences of diminishing local populations are much less catastrophic than for species since mating with individuals from other subdivisions of the species is possible. In fact, in many cases, much of human 'migration' is simply exogamy with marriage partners being found in neighboring groups (recall the Gainj described in Chapter 2). One cause of migration, then, is the tendency for outmarriage found in many cultures, a pattern presumably of great antiquity in the human species (Tylor 1888), and one affected by numerous causal factors itself.

The degree of endogamy (or its converse, marital migration rate) is a key component of population structure. Low endogamy (high gene flow) leads to reduced local inbreeding and less genetic differentiation (although as we have seen, high rates of migration do not necessarily imply genetic homogenization; Fix 1978). Isolation has traditionally been seen as a prerequisite for speciation (Mayr 1963) and, all other things being equal, the greater the degree of endogamy, the less likely is it for gene flow to overcome the differentiating effects of genetic drift. Similarly, the long-term importance of founder effect (or lineal fission; Neel & Salzano 1967) depends on the founding group establishing itself as a viable population, otherwise, the founding gene pool would be swamped by later immigrant spouses from other groups. Recent representations of human evolution as a branching process also imply fission of groups, migration to new territories, and relative isolation from other populations (Cavalli-Sforza *et al.* 1994). While such tree representations are literally wrong, since no human population has been reproductively isolated from the species, the *degree* to which fission groups could maintain relative isolation is a crucial factor in the development of genetic differences among groups.

The potential for small populations to be endogamous depends on numerous factors. As Adams & Kasakoff (1976) have pointed out, endogamous groups are basic units for social interaction as well as being genetic units and therefore any factor limiting or extending social contact may influence endogamy. Many of these factors have already been considered in previous chapters. Geographic isolation such as in the case of the island of Tristan da Cunha (Roberts 1967), located in the midst of a vast sea with little contact with other populations, can severely constrain the possibility of marrying out. Conversely, economic factors such as trade can extend the social universe of local populations and encourage outmarriage. Demographic structure of the local population can affect the numbers of potential mates and thereby endogamy rates (MacCluer & Dyke 1976). Often seen as 'regulating' marriage, incest taboos and cultural prescriptions to marry certain categories of kin or others can also affect the potential for endogamy (Hammel *et al.* 1979; Kunstadter *et al.* 1963; MacCluer 1974; Morgan 1974).

Complex phenomena depending on the interaction of many determinants are good candidates for experimentation via computer simulation. Analytic models of migration or population structure have difficulty incorporating such complexity. As Ewens *et al.* (1987:60) point out in discussing values for minimum viable population size estimates, 'theoretical calculations may be possible only for unduly simplified models'. They recommend computer simulation as the most useful approach to this problem.

More directly relevant to the study of potential endogamy, Dyke (1981:195) notes that algebraic models in demography do not include a suitable function for marriage and traditional demographic approaches do not consider the effects of random processes in small populations. Both of these deficiencies can be corrected via simulation. The computer can be programmed to simulate a marriage system each attribute of which can be varied to ascertain the effects on the potential for endogamy of the population. Random number generators allow stochastic or Monte Carlo simulation of random variation. Although these experiments do not provide an analytical solution to the problem, they do offer insights based on the more realistic population structures.

Simulation studies of marriage systems have a relatively long genealogy in anthropology. The first use of simulation in anthropology was probably that of Kunstadter *et al.* (1963) who looked at the potential for realizing ideal marriage rules in the face of the expected variability in the availability of preferred marriage partners in small populations. Their work was followed by a series of studies of mating systems and demography (reviewed in Dyke 1981), several of which were concerned with potential endogamy under varying demographic and cultural regimes (e.g., Fix 1982a; Hammel *et al.* 1980; MacCluer & Dyke 1976).

These studies emphasize the effect of demographic and social structure as causes of exogamy and gene flow. The majority of causal analyses of migration discussed in Chapter 1 focused on economic push/pull factors. However, the presence or absence of potential spouses may be the critical factor in a decision to migrate in many societies. Models and analyses of these intrinsic demographic and cultural factors affecting movement can help to explain patterns of mobility in these populations.

The usual prerequisite for reproduction in human populations is marriage. In the absence of a suitable mate, individuals must remain celibate or emigrate in search of a spouse (Leslie *et al.* 1980). The cultural prescriptions and proscriptions in terms of kinship and age defining a suitable mate interact with the kin and age structure of the population to delimit the potential mate pool of an individual (Dyke 1971). Such pools may be empirically estimated from the actual composition of the population (Dyke

1971), however, simulation provides a useful technique to model the affect of demography and rules on potential mate pool size.

The population structure and basic social organization of the Semai Senoi of Peninsular Malaysia was described in Chapter 2. Considerable data exist on Semai demography and marriage patterns (Fix 1977) allowing many parameters of their marriage system to be specified. Part of the demographic study of their population included Monte Carlo computer simulation of variation in vital rates and processes (Fix 1977). This demographic simulation can be used to address the question of the potential for endogamy in Semai settlements (Fix 1982a).

Recall that Semai settlements are generally small, ranging from 25 to more than 270 persons. Semai have no formal rule regarding settlement endogamy or exogamy. There is, however, a general mistrust and fear of 'strangers' (Dentan 1968) and most Semai would prefer to marry someone they know or know about. This preference translates into a tendency to marry endogamously when possible and often into a family already connected by marriage (Benjamin, 1986 notes this preference for marrying affines among other Senoi of Peninsular Malaysia).

A stronger rule is that prohibiting marriage between close kin. Semai say this proscription includes *all* relatives but it seems to apply usually to kin within the range of second cousins (in a group of 129 marriages for which genealogical information was sufficient to gauge, no first cousin and only nine second cousin marriages occurred; Fix 1982a). Since relatives are often localized in the settlement of residence, and the smaller the population, the greater the likelihood of a potential spouse being kin, this rule may have a strong affect on endogamy. The very small, very isolated population of Tristan da Cunha nicely illustrates this principle. As the generations from founding passed, the only potential mates available were relatives (Roberts 1967). Among the Semai, the strong sense that consanguineal kin should avoid marrying combined with an ideology of kin solidarity leads to an apparent conflict. Coresidents often express their unity by stating, 'we are all kin here', by inference, all the members of the local group will cooperate and help each other as would kin. Syllogistically, if *all* members are truly kin, then *no* member can marry any other and the group must be exogamous. When presented to them in this way, some Semai agreed that local groups were exogamous. Actual data on marriages (Table 2.2), however, showed some 45 percent of spouses were both born in the same settlement and a greater number were coresident at the time of marriage. The endogamy rate in any settlement, then, depends on the presence of non-kin in local groups. The fission–fusion structure of Semai settlement histories usually ensures that some more distantly related persons will be available within a settlement.

A further constraint on potential spouses is relative age difference. For nearly all child-producing marriages, males are between 20 years older and 5 years younger than their female spouses (Fix 1982a). For most, the range is narrower, males being within 10 years older to 5 years younger. Semai polygyny is so rare as to be ignorable and the great age disparities between spouses associated with this practice are not found.

Essentially all Semai women greater than 16 years old are married, the few exceptions include recently widowed or divorced women (who quickly remarry) or sterile women, a few of whom remain unmarried. For men, the problem of getting and staying married is more difficult. Formerly the sex ratio among adult Semai was biased in favor of men (Fix 1977). The overall sex ratio was considerably greater than 1.0 and, for the study area described in Chapter 2, was 1.27. This imbalance was apparently due to high maternal mortality rates differentially affecting women in the reproductive years. Recent improvements in maternal care have reduced the number of maternal deaths and the sex ratio is now becoming more even (Fix 1991). In the past, however, unmarried men of 25 or more years old were not uncommon and some older men were also single. Most men do find mates, however, and the proportion of adult single men is only about 12 percent.

To summarize, a Semai preference to marry endogamously is constrained by the proscription on marrying near kin, age preferences, and, in the case, of men, a relative shortage of women progressively becoming more acute through the reproductive years. The question is, what is the minimum population size that would allow Semai to marry endogamously given their marriage system and demography? A secondary question concerns the degree to which marriage rules might affect this potential, particularly the effect of relaxing cousin avoidance rules.

The requirements of a simulation model to explore these questions include an adequate representation of Semai demography (fertility and mortality) and a method of simulating Semai marriage. The basic characteristics of a FORTRAN program that satisfies these conditions is shown in Table 4.1 (see also Fix 1977, 1982a). In brief, the artificial population is constructed from an input population of individuals representing the population to be tested for potential endogamy. In these experiments, various settlement populations were chosen as initial groups ranging from the largest Semai settlement in the region, with a population size of 272, to one of the smaller settlements ($n = 50$). Another set of experiments was done with a fission group (a complex set of 51 interrelated individuals).

Table 4.1. *Stochastic computer microsimulation of Semai marriage*

Input data
Information on each individual in the initial population including:
Age, sex, unique identification number (ID), father's ID, mother's ID, spouse's ID
Age and sex specific probabilities of death
Age specific probabilities of giving birth
Order of operations (repeated for the number of years specified)
Mortality
Generate random number $0 \geq x \leq 1.0$
Compare to age-sex specific probability of dying (ASPD)
If $x <$ ASPD, record death and remove individual from array
Mating
Unmarried males matched with unmarried females in appropriate age categories
If potential spouses kin (degree specified as input), reject marriage
If appropriate mating, record marriage
Birth
Married females give birth
Using similar procedure to mortality except x compared to age specific probability of birthing
Output
Basic population statistics including births, deaths, and associated rates, population size
Matings, matings rejected due to relatedness

The initial population comprised four arrays of individuals, divided into married and unmarried males and females. For each individual, information on age, a unique identification number (ID), and the IDs of father, mother, and (if married), spouse was stored. These ID numbers allowed the degree of relationship of potential mates to be assessed. As the simulation proceeded, year by year, this original population gradually was reduced by mortality until at the end of 100 or 200 years none remained.

Basic demographic parameters including age and sex specific probabilities of dying, and age specific probabilities of giving birth, were also part of the original input for the simulation. These rates of birth and death were applied each year to the population. Because these are very small populations, stochastic variation in these rates is pronounced. In order to simulate this random variation, both mortality and fertility were simulated by a Monte Carlo technique. Each individual each year had a probability of dying based on the age and sex specific death rates (ASPD) appropriate to him or her. A random number, x, between 0 and 1 was generated for each individual. If x was less than the ASPD, that individual was recorded as a death and removed from the array. This procedure ensured that the average death rate in the simulated population would be the same as that derived from the real population but it also built in stochastic variation around that average.

Following the mortality procedure, each unmarried male in the appropriate age categories was matched to the list of unmarried females. This process will be described in more detail below; for now, it may be noted that marriage was a prerequisite for reproduction. The decision whether a woman would give birth was simulated in the same way as for mortality. The yearly age-specific probability of reproducing was compared to a random number, and a new birth (male or female determined by chance) was recorded and stored with the current year as birthdate and with mother's and father's genealogies appropriately appended. These procedures were repeated for the number of years specified for each experiment, usually 100 to 200 years. At intervals of five years, demographic statistics including births, deaths, age and sex composition of the population, number of matings, and summary rates and ratios were printed.

To validate the program, these statistics were compared with expectations from a model population fit to the Semai data (Fix 1977). Since the age–sex specific probabilities of dying in the simulation were identical to the model values, the averages of many events in the artificial simulated population ought to be the same as the model. Table 4.2 shows that a close match was achieved. Other measures such as the age composition of the population and the crude birth and death rates also show good agreement with the model. Crude birth and death rates in the artificial population averaged 41 per thousand and 34 per thousand respectively compared to 42 per thousand and 35 per thousand in the stable model. These results provide confidence that the simulation model is correctly mimicking the basic demographic processes of the Semai population.

To analyze the endogamy potential of these populations, all mating was confined to the local group. If a male was unable to find a suitable mate in the search through the list of unmarried females, he remained unmarried. Ultimately, if too many persons were unable to mate and reproduce, the population might fail to be perpetuated. This is, of course, the idea behind the minimum viable population concept of conservation biology (Soulé 1987). In the Semai case, and it may be assumed for all but the most severely isolated human populations, individuals unable to find mates locally would search elsewhere. Nonetheless, a convenient measure of the potential for endogamy is the percentage of the population failing to find mates in the local group. Recalling that the high sex ratio in the real Semai population means that some 10 percent of the males may be unmarried even with the option of finding mates in other settlements, a realistic expectation for endogamy might be for all females to be married but not all males.

Table 4.2. *Comparison of artificial and model populations*

Variable	Artificial[1]	Model
Infant death rate	0.196	0.200
Sex ratio at birth	1.070	1.068
Female l_{45}/l_{15}	46.3	46.0
Completed family size	5.72	5.73
Annual rate of growth	0.0066	0.0070

Source: Fix (1982a).
[1] Average rates from five runs of 200 years.

While the mortality and fertility components of the simulation remained constant throughout the series of experiments, two different conditions for kin exclusion in the mate choice component were employed. The first implemented the fairly strict rule that excluded second cousins as potential mates. The Semai claim that no relative is marriageable but genealogical data show that although no marriages of first cousins occurred, nearly 7 percent of marriages were with second cousins. Mate choice under the second condition allowed much more latitude; only first cousins and closer kin were prohibited. Age preference was accommodated under both conditions by searching in the mate seeker's age category of eligible women first, then one age category (5 years) below, one above, two below, three below, until the lower limit of female mating age (15–19 years) was reached. Each male searched until mated or until the list of his potential mates was exhausted. Widows and widowers also entered the mate pool upon the death of their spouse and could be remarried so as not to inflate the total number of unmarried persons.

Ten replicate runs each were made of three different initial populations for each of the two conditions of kin exclusion for a total of 60 runs. Table 4.3 and Figure 4.1 show the results of these experiments. The averages of the 10 runs for each condition are represented here. Some individual runs showed wide chance deviations from these averages, the sort of stochastic variation one would expect among individual settlement populations. Averages of final population sizes (N_f) are after 100 years of simulation, the annual growth rates are again averages over 10 replicate runs over the entire 100 year period of the simulation. The proportions of unmarried males and females are averages for the last 50 years of the simulation after most of the original input population had been replaced.

These experiments suggest that large Semai settlements might be able to marry endogamously over a relatively long period and still avoid mating with close consanguineous kin (Condition 1). The large population

Table 4.3. *Proportions of unmarried individuals in the three artificial populations*[1]

| Population | Condition | N_i | N_f | r | Males | | Females | |
					%	range	%	range
Large	1	272	459	0.0052	0.070	0.004–0.159	0.027	0–0.072
Large	2	272	497	0.0060	0.058	0–0.168	0.078	0–0.186
Small	1	50	47	−0.0006	0.478	0.355–0.638	0.303	0.114–0.600
Small	2	50	65	0.0027	0.223	0.039–0.511	0.085	0–0.216
Fission	1	51	47	−0.0009	0.486	0.257–0.810	0.365	0.200–0.570
Fission	2	51	75	0.0038	0.215	0.051–0.621	0.209	0.038–0.548

[1] Each of the three populations, representing a Large settlement, a Small settlement, and a Fission Group were simulated for both consanguineous mating exclusion conditions. Condition 1 forbade matings with all relatives through second cousins; Condition 2 allowed second cousin but no closer kin. Averages are for the 10 runs for each condition. Proportions unmarried are averages over the last 50 years of the 100 year runs; the ranges of percentages unmarried also represent this period. N_i is the initial input population size; N_f is the final population size after 100 years; r is the annual rate of growth (N_f and r are averages of 10 runs).

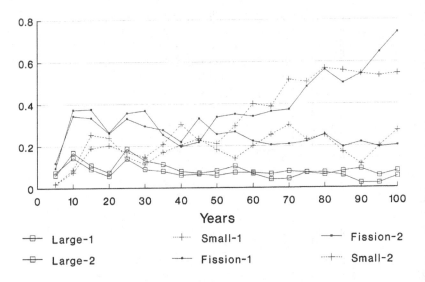

Figure 4.1. Comparison of cumulative mean percentages of unmarried males (within the 20–44 age class). (From Fix 1982a, fig. 4.)

($N_i = 272$) nearly doubled in 100 years under the stringent avoidance rule and only a few persons were unable to find mates. It should perhaps be noted that real Semai might have more conditions to place on a potential spouse than the computer that searched relentlessly for an unrelated person. However, the simulation results do show that *if necessary* a relatively high degree of endogamy is possible under this demographic and cultural regime. Under Condition 2 (second cousin mating allowed), males showed a slightly lower average of unmarrieds and the overall growth of the population was a bit greater. The proportion of second cousin marriages under Condition 2 was considerably greater than in the actual population averaging between 10 and 20 percent compared to about 7 percent.

In contrast to the large initial population, both the small and fission group populations declined in numbers under Condition 1 and many males were unable to find mates. Three runs with initial small population size were extended to 200 years and all three were declining to extinction by the end of the run. Nearly half the males and a substantial portion of the females were unable to find mates under these conditions. Figure 4.1 shows graphically how the percentage of unmarried males builds up over time in the small populations.

Another measure of the difficulty of finding mates in these small populations is the high number of matings that were rejected due to relatedness. In the large population, the average number of matings during the last five years of the runs was 150. Another 153 potential pairings were rejected since their relationship was closer than second cousin. This may be compared to the small fission group where during the same time period only six matings were allowed and 1769 potential pairings were rejected. Clearly in the fission group initial kinship had built up over time to very high levels of overall relatedness – essentially *everyone* was related too closely to marry.

When the potential mate pool was enlarged by including second cousins (Condition 2), survival was possible but population growth was limited and many males and females remained unmated. In the fission group, over one-fifth of males and females were single over the last 50 years of the simulation. The build up of relatedness in the population is apparent in the graph of second cousin matings (Figure 4.2). The cumulative average percent of matings that were second cousins increased steadily in both small populations reaching some 70 percent in the fission group. The average in the larger population, while high by Semai standards, remains lower and fairly constant through time.

The initial size of the population has been shown to be critical in determining the potential for endogamy. Populations experiencing Semai-like demographic regimes (moderate to high mortality and fertility) would be unlikely to possess endogamous units below 100 persons for reasons intrinsic to the population. Since the results presented here have been *averages* of outcomes, even larger groups might by chance fail to maintain endogamy.

Because these are experiments employing numerical values for the parameters of the model, the results are not generalizable in the same way as an algebraic or analytical model. In so far as other populations might differ from the Semai conditions modeled here, the results might also differ. The way around this limitation is for further experimentation using different parameter values (or even different algorithms) to both further validate the model results and extend the range of conditions for which these results hold (Dyke 1981).

For the problem considered here, potential endogamy, several other simulation studies have explored various aspects of demography and marriage rules allowing broader conclusions to be drawn than from the single set of experiments. Hammel *et al.* (1979), for instance, confirmed that incest rules were of much greater importance in small (dozens of persons) closed populations compared to large (several hundred) populations. This parallels the finding that exclusion of second cousins in the large Semai population had relatively little effect on the survival of the population. MacCluer & Dyke (1976) examined both mating pattern and fertility and

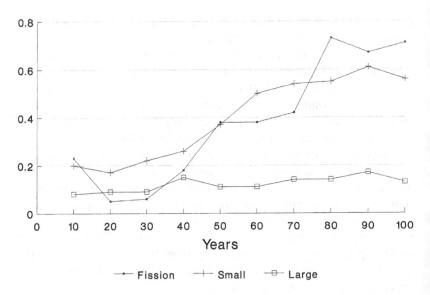

─•─ Fission   ─+─ Small   ─□─ Large

Figure 4.2. Comparison of cumulative means of second-cousin matings as percentages of total matings. (From Fix 1982a, fig. 5.)

mortality levels in their simulation study of minimum endogamous population size. They found that populations of one or two hundred persons could maintain endogamy. They systematically varied mortality and fertility rates and discovered that populations with high rates are most influenced by random fluctuations and therefore require larger population sizes to maintain endogamy. On the other hand, kin exclusion rules most influence populations with low mortality and fertility since biological relatedness is higher in these populations. The strictest rule that they considered prohibited first cousin marriage. Semai avoid a wider circle of relatives, a factor that increased their minimum mate pool size. MacCluer & Dyke (1976:11) conclude that most additional factors contributing to endogamy potential, including political, economic, or sociological factors, would tend to increase the minimum viable size of groups such that 'it would be reasonable to question whether a real population which survives for any length of time with fewer than one hundred individuals is truly endogamous'.

Interaction of genetic drift and migration – kin-structured migration (KSM)

The previous section has shown how a complex demographic process can be dissected into components and simulated numerically on the computer.

Just as these demographic events such as mating, birth, and death can be simulated, the forces of genetic evolution are also amenable to simulation modeling.

The initial demonstration of the role that KSM could play in genetic microdifferentiation was based on a simulation model (Fix 1978). The migration model used to explore this problem considered three factors not present in the algebraic models discussed in Chapter 3: (1) the migration of groups of individuals; (2) the kin-structuring of these groups; and (3) a random element in their movement. Since this work was completed, Rogers (1987), as noted above, has devised an analytical model of KSM that allows a more general treatment. This history represents a good example, I think, of the initial use of simulation to experiment in situations where no algebraic models yet exist thereby stimulating the development of a more general approach.

Table 4.4 presents the components of the simulation program. Gene frequency change in a linear series of 25 populations, each of size 250, was simulated for 20 generations for each run. Since the problem of interest was the effect of KSM on genetic differentiation, all runs began with equal gene frequencies in all populations. Further, since balancing natural selection acted as a stabilizing force to counteract genetic drift in several of the experiments, initial gene frequencies were set at the equilibrium frequency determined by genotype fitnesses. In these runs, the equilibrium value of q was 0.25. Because the model is stochastic and numerical, several replicate runs must be made to gauge the amount of variation; for these experiments, five repetitions were made under each experimental condition.

The simulation sequence began each generation with gene frequencies in each population in the array being randomly determined according to the expectation under genetic drift – i.e., from a normal approximation to a binomial distribution with the expected value being the current generation's gene frequency and variance equal to $pq/2N$. Recalling the discussion from Chapter 3 on the life cycle at which migration takes place, it should be noted that this sequence presumes that population regulation and genetic drift occur *before* migration, a more realistic assumption for humans and, indeed, most animals.

Following drift, gene frequencies for each population are again recalculated according to the formula for balanced selection (shown in Table 4.4). To continue the life cycle issue, selection occurs in this model among the children, the survivors of whom migrate as young adults. The fitnesses were chosen to approximate those determining the hemoglobin A/E polymorphism in Southeast Asia and represent strong selection (Fix 1978).

Table 4.4. *Kin-structured migration computer simulation model*

Order of Operations (repeated for the specified number of generations):
Genetic drift
    Randomly determine gene frequencies of each population
Natural selection
    Determine gene frequencies after balancing selection according to

$$q'_i = \frac{q_i(w_2 p_i + w_3 q_i)}{q_i(w_2 p_i + w_3 q_i) + p_i(w_1 p_i + w_2 q_i)}$$

Deterministic migration
    Apportion migrants between populations according to

$$q'_i = (1 - m)q_i + 0.4mq_{i-1} + 0.4mq_{i+1} + 0.1mq_{i-2} + 0.1mq_{i+2}$$

Kin-structured migration
    Randomly determine gene frequencies of two migrant groups from each population
    Randomly apportion migrant groups to adjacent populations

---

Uniform balancing selection should counteract the differentiating force of random drift in these relatively small populations.

A further stabilizing force is migration, represented here as a deterministic formula (see Table 4.4). This formula is a modified linear stepping-stone model, modified in that exchange is not only with immediate neighbors but extends to two populations on either side of the donating group (this extension follows Livingstone 1969). The total amount of migration in each generation varied from 0.20 in those runs without kin-structured group migration to 0.06 in those runs with kin-structure.

KSM was simulated by choosing two migrant groups from each population each generation. Gene frequencies in these migrant groups were determined randomly from a normal distribution with the expected value being the gene frequency of the donor population and with variance $pq/2AN_m$, where N_m was the number of migrants and A was a reduction factor varying with the degree of relatedness among the migrants. Depending on the degree of kinship among the group, the sample of migrants may represent many fewer independent genomes than their actual census number. Thus a group of siblings can carry only copies of the four genes at each locus present in their parents.

Compared to the deterministic migration equation, this procedure adds a random element to migration. Migrants are viewed as a *sample* of the parental gene pool and, additionally, kin-structured groups are a *biased* sample since each element is not independent due to the probability of sharing genes by descent. The similarity of random migration to genetic drift is clear in this model from the equivalent algorithm used to choose gene frequencies for both processes. As Rogers (1988) noted elsewhere, this form of migration is actually better conceived of as one component of genetic drift.

A further source of randomness under kin-structured group migration is the *destination* of migrants. As has been discussed in Chapters 2 and 3, fission–fusion societies such as the Yanomamo or the Semai, periodically split, and groups fuse with neighboring villages. Which village they choose to join is often affected by ties of kinship or previous history and politics, and is not truly random. However, in the model developed here, the probability of fusion groups moving to one of the four adjacent neighboring populations was set at the same frequencies as deterministic migration – i. e., 80 percent chance of going to one of the nearest neighbors and 40 percent of moving two populations in either direction.

One group of 10 individuals represented a set of close relatives such that only about half of the actual number of genes present in the group were independent ($A = 0.5$ and $2N_m A = 10$). Another larger more dilute kin group ($n = 25$) was also chosen such that $A = 0.75$. After migrant gene frequencies were randomly chosen, a recipient population from the four populations on either side of the donor population was randomly selected. The gene frequency of the migrant group (in proportion to the total population size) was added to the recipient population and subtracted from the donor population. This latter step may also have important implications for gene frequency variation in small populations. Consider particularly rare alleles that might be concentrated in families. Emigration by a family group may remove the allele from the population (recall the case of Tristan da Cunha from Chapter 3; Roberts 1968).

Each module of the program is potentially self-contained, therefore, different combinations of the procedures may be combined in order to test for the effect of each process and/or different parameter values used within a procedure. A variety of experiments were performed using this model, but the role of kin-structure can be clearly shown by comparing four run sets. Table 4.5 summarizes these sets. Figure 4.3 presents the normed genetic variance for each run set over the course of the 20 generations of the runs (these values are means calculated from the five replicate runs for each set).

As Figure 4.3 makes clear, three of the run sets conform to expectation, that is, the genetic variance generated among the populations is balanced by selection and migration and stabilized through time. Strong uniform balancing selection and a high rate of migration (run set 1) keep variation among the populations at a fairly low level. When individual (deterministic) migration is augmented by random groups either not kin-structured (run set 2) or kin-structured (run set 3), higher levels of variation are

Table 4.5. *Summary of run sets*

Run set	Conditions	Deterministic m	Total m
1	RGD, SEL, DM	0.20	0.20
2	RGD, SEL, DM, KM 100–100	0.06	0.20
3	RGD, SEL, DM, KM 75–50	0.06	0.20
4	RGD, DM, KM 75–50	0.06	0.20

RGD: Random genetic drift; SEL: balanced selection; DM: deterministic migration; KM: Kin-structured migration with A values for the two groups (100–100 indicates no kin-structure).

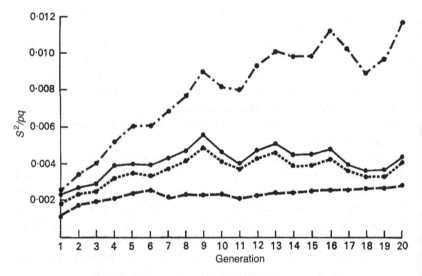

Figure 4.3. Mean values of the standardized genetic variance under four conditions. All four experiments included random genetic drift and deterministic migration. Differences included: (1) balancing selection (- - - -); (2) balancing selection, and 'founder' migration (.....); (3) balancing selection and kin-structured migration (——); and (4) kin-structured migration (— · —). (From Fix 1978, fig. 3.)

maintained but there is no trend for increasing levels through time. Run set 4, however, is particularly instructive. In the absence of strong selection, but with a high rate of total migration (20% per generation), genetic variation rises continuously over the course of the runs. This steadily increasing variance was verified by extending the run to 100 generations.

These simulation models allow more realistic models of migration involving random components to be investigated. For several societies treated in Chapter 2, migration is structured along kin lines, these more complex models may give us a better expectation for genetic variation than the classic population structure models.

Migration–selection interaction – detecting clinal and balanced selection under KSM

The simulation analysis in the preceding section showed that KSM is less effective at reducing variation generated by random genetic drift and, in fact, may contribute to such variation under certain circumstances. These findings suggested that attempts to infer the operation of specific evolutionary processes from patterns of genetic variation might be subject to an additional source of uncertainty, the 'extra' variance introduced by random KSM (Fix 1994).

Inference of process or history from observed patterns in gene frequencies is the stock in trade of much of population genetics. In the next chapter, we will see how entire historical scenarios are created from gene frequency data (most notably in the massive compendium of Cavalli-Sforza *et al.* 1994). At the same time, the major issue for population genetics theory through much of the 1960s and into the 1970s was focused on discerning the relative roles of random and selective forces in evolution, the so-called 'neutralist-selectionist' debate (Lewontin 1974). Although many would now claim that this issue has been resolved in favor of the neutral theory (at least with regard to molecular genetic variation), it would still be useful to be able to recognize (even if rare) cases in which allele frequencies have been affected by natural selection.

In theory, such discrimination should be possible. The key difference between natural selection and genetic drift/migration is that selection should be locus-specific in its mode of operation whereas the other two forces should act uniformly over all loci. Thus particular alleles are targets of natural selection while drift and flow result from population parameters, size and migration rate, that affect the entire genome. However, it might be noted parenthetically that the list of human genes having been shown to resist malaria has reached an impressive length now including several hemoglobin types, enzyme and red cell membrane 'defects' (Livingstone 1985) and even an HLA type (Hill 1991) suggesting that powerful selective agents such as endemic or pandemic disease might act on many loci in concert.

Lewontin and Krakauer (1973) applied this theory by examining heterogeneity among alleles in Wright's F_{ST} to detect natural selection. The problem with their test was the lack of a suitable statistical criterion for detecting significant departures from expectation (Felsenstein 1982). More recently, Sokal *et al.* (1989b) have used a different statistical procedure, spatial autocorrelation analysis, to detect selective effects based on the same underlying theory of locus specificity.

Spatial autocorrelation analysis plots the correlation between an allele frequency in one population with the frequency of the same allele (whence the 'auto') in another population and arrays these correlations by distance class in correlograms (Cliff & Ord 1981). These plots should show characteristic signatures dependent on the underlying process that generated the genetic variability they record. Thus Sokal & Wartenburg (1983) showed that for neutral alleles under isolation by distance in a continuous population model, correlograms should describe a steep decline with distance class (very much like the negative exponential decline of genetic similarity with distance seen in the Malécot model described in Chapter 3). In contrast, alleles undergoing selection should show patterns of spatial autocorrelation consistent with the type of selection and distinctively different from the neutral isolation by distance curve. Clinal selection should produce correlograms with a linear decline from positive autocorrelation at short distance to negative at long distance. Uniform selection across all populations should result in no spatial pattern of autocorrelation. These are quite distinct predictions and, in theory, should allow the detection of the evolutionary force from the gene frequency data.

The ability of spatial autocorrelation analysis, and for that matter any statistical method, to detect departures from neutrality is dependent on the strength of the evolutionary force (the signal) and the stochastic variation generated by population structure (random noise). If directional or clinal selection is weak, it will be impossible to distinguish selected from neutral loci amidst the background of genetic variation due to drift or other random processes. The greater the noise, the more difficult the detection of the signal. Thus, even strong selection could remain undetected if the population structure produces a great deal of random variation.

Nearly all the forces of evolution are capable of increasing genetic variation. Ordinarily the set of factors included in the category genetic drift are principally implicated in generating variation while the systematic forces (migration and selection) stabilize gene frequencies. However, as Wright (1948) noted in his catalog of modes of gene frequency change, 'fluctuations in the systematic pressures' can also increase variation.

Intergenerational genetic drift is the usual exemplar of the category. As Rogers (1988) has pointed out, there are several other components of genetic drift in subdivided populations. Founder effect (Mayr 1963) is a powerful differentiating force, but even more effectual is the kin-structured variant called lineal effect by Neel and his colleagues (Neel & Salzano 1967;

Smouse *et al.* 1981). Also capable of augmenting genetic variation is stochastic migration and once again the kin-structured form of random migration increases the effect.

A further consideration when studying empirical gene frequency distributions rather than theoretical ones is the problem of sampling error in calculating gene frequencies (Epperson 1993). Thus the detection of specific evolutionary forces from gene frequency distributions is complicated by the numerous sources of genetic variation, and the full range of these factors needs to be examined (Slatkin & Arter 1991).

As discussed in Chapter 3 and in the previous section, KSM may have substantial effects on genetic variation. We now turn to a computer simulation study that evaluates these effects on the ability to discern natural selection from spatial correlograms (Fix 1994).

The simulation model used to explore this question was a modification of the basic program already described in the preceding section (see Table 4.4 for a summary). A linear stepping-stone array of 25 populations exchanging kin-structured migrant groups between nearest neighbors only (one-step in contrast to the two adjacent populations on either side in the previous model). Population size (n) was set at 100 (some experiments were carried out with ns of 50 and 200). The simulations were run for 50 generations and most were replicated 10 times with different random number initializers.

Genetic drift was simulated in an identical fashion to the previous model by randomly determining gene frequencies for each generation as a sample from a normal distribution with the mean being the old gene frequency and with binomial variance. All populations began with identical initial gene frequencies for 8 di-allelic loci ($q = 0.5$, except for the balanced selection case where $q = 0.2$, the equilibrium frequency).

Three different selection regimes were considered: (1) directional selection against a recessive; (2) balanced; and (3) clinal selection (see Table 4.6). Directional selection comprised 10 percent selection against a homozygous recessive genotype. For the balanced case, selection occurred against both homozygous genotypes such that a balance at $q = 0.2$ was achieved. Selection was modeled by the formula previously given (Table 4.4), i.e.,

$$q'_i = \frac{q_i(w_2 p_i + w_3 q_i)}{q_i(w_2 p_i + w_3 q_i) + p_i(w_1 p_i + w_2 q_i)}$$

Clinal selection was simulated by varying the selective coefficient against a recessive homozygote according to the formula

$$s = 0.01d$$

Table 4.6. *Simulation models*

Model	Selection regime	N	m (%)
KS	Neutral	100	10
DS	Directional ($w_1 = w_2 = 1.0; w_3 = 0.9$)	100	10
DSS	Directional ($w_1 = w_2 = 1.0; w_3 = 0.9$)	50	20
BS	Balanced ($w_1 = 0.875, w_2 = 1.0; w_3 = 0.5$)	100	10
CS	Clinal ($w_1 = w_2 = 1.0; w_3 = 0.99$ to 0.75)	100	10

See text for details of models. N is the population size; m is the migration rate; the $w_{(1-3)}$ are the fitnesses of the three genotypes.

where d is the population location in a linear array from 1 to 25, and thus s varies from 0.01 to 0.25. The correlograms produced by these selection schemes are compared to those generated by neutral loci under isolation by distance.

Migration of sets of siblings was simulated ('kin-structured' in this case is actually 'sib-structured' migration). The procedure to simulate sibling group migration entailed randomly choosing the genotypes of two 'parents' from the donor population, then, based on these parental genotypes (for each locus), generating the genotypes of five offspring. Depending on population size, m was 10 percent (10 migrants from the populations of 100, five sibs going to one adjacent population and five to the other) or 20 percent (10 migrants from populations of 50). This model simulates linear gene flow which is not only stochastic in that migrants are samples of their natal populations but are additionally kin-biased samples. This model differs from the previous kin-structured simulation in that the destinations of the migrants are predetermined. Instead of being randomly apportioned to one of four possible neighboring populations, each neighbor receives one sibship.

Following migration, spatial autocorrelation coefficients were calculated for the system. The measure of spatial effects on gene frequencies used here is Moran's I, one of several measures of spatial autocorrelation (Cliff and Ord 1981) popularized by Sokal and his colleagues (e.g., Sokal and Wartenberg 1983). Populations are grouped by distance class and the correlations between gene frequencies for each locus in each class, k, are computed by

$$I_K = n \sum_{ij} w_{ij}(q_i - \bar{q})(q_j - \bar{q})/W_k \sum_{i=1}^{n} (q_i - \bar{q})^2$$

where n is the number of populations, $\bar{q}$ is the average gene frequency for the allele over all populations, $w_{ij} = 1$ for pairs of populations in the same distance class k and 0 otherwise, and W_k is twice the total number of pairs in distance class k (see also Epperson 1990 for notation). The values of I for each allele k are plotted by distance class to form a correlogram.

The results of these experiments show that the detection of natural selection amidst the random background of the dispersive forces is difficult under the conditions modeled here. When relatively small populations ($n = 50$, 100, or 200) are the units of analysis, gene frequency variation generated by the stochastic forces of drift and kin-structured migration often obscures the characteristic signature correlograms expected under the various forms of natural selection.

Figures 4.4 through 4.6 present the means and two standard deviation error bars for 10 alleles from 10 replicate runs under selective conditions DS, BS, and CS (directional, balanced, and clinal selection, respectively; see Table 4.6). Since for each run, half the loci were subjected to selection and half were not, each point in the figure represents the mean of 50 values of Moran's I (5 loci × 10 runs), likewise, the error bars are based on standard deviations calculated on the basis of the same 50 values. The solid line represents selected loci; the dashed line, neutral loci.

All three figures show that average autocorrelations for neutral loci decline swiftly from the first distance class ($k = 1$, average Moran's $I = 0.445$) to nearly 0 by distance class 3. Beyond $k = 3$, little spatial structure is apparent. The error bars for neutral loci are quite wide, extending from near 0 to greater than 0.8 for $k = 1$ (see also Fix 1993, Fig. 7). In the absence of selection, isolation by distance produces a characteristic decline of spatial autocorrelation with distance (Sokal and Wartenburg 1983) as an expectation, however, considerable variation is apparent in these simulated data.

Figure 4.4 (DS) compares correlograms of alleles subject to directional selection with those produced by neutral alleles. The expectation under directional selection against a recessive homozygote is asymptotic loss of the selected allele. The spatial expectation is for reduced autocorrelation (Epperson 1990) since selection affects all populations uniformly. Thus, spatial autocorrelations generated by isolation by distance should be reduced (and ultimately, as the allele is lost, all spatial structure is lost).

Interestingly, these simulation results demonstrate that even under relatively strong selection ($s = 10\%$), the dispersive forces of drift and KSM are sufficient to retain much variation among correlograms. The error bars of the selected loci nearly completely overlap those of the neutral loci. Moreover, the means of Moran's I are nearly identical for the distance classes greater than 3.

Figure 4.5 (BS) contrasts correlograms of loci under balanced selection

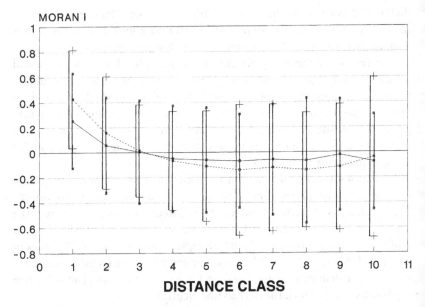

Figure 4.4. Average correlograms (10 replicate runs), directional selection against a recessive (solid line) and neutral (dashed line); the error bars extend two standard deviations from the mean (small squares for selected, + for neutral loci). (From Fix 1994, fig. 1.)

and neutral loci. As for the directional (DS) case, selection may be expected to reduce spatial autocorrelation structure since all populations along the linear array are subject to a uniform force. The selective values modeled are very great (50% against one homozygote; 12.5% against the other) and are strong enough to reduce average autocorrelations across all distances to nearly 0. Despite this strong, effective selection, variation persists in autocorrelation structure. Again, as for the previous condition, mean values of Moran's I for $k > 3$ are not very different between selected and neutral loci and the error bars nearly completely overlap.

Figure 4.6 (CS) shows that strong clinal selection (increasing from 1% to 25% in increments of 1% across the 25 populations) produces a characteristic signature correlogram as the average or expectation of the process. More relevant to the goal of distinguishing such underlying processes, the average clinal correlogram exhibits high positive autocorrelations for several distance classes beyond the first 3, becoming negative only at $k = 9$. This pattern contrasts with neutral correlograms and with the other selective regimes as well. On the other hand, the error bars continue to be broad and the mean of the selected loci is included

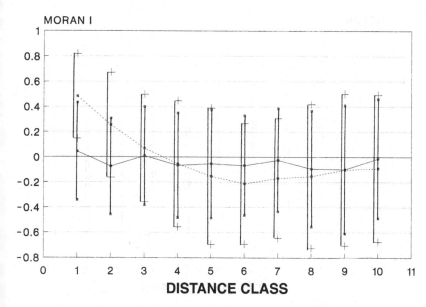

Figure 4.5. Average correlograms, balanced selection (solid line) and neutral (dashed line). (From Fix 1994, fig. 2.)

within the interval for the neutral loci in all but two of the distance classes ($k = 3$ and 4).

Clearly it would be useful to be able to assess the statistical significance of these differences among correlograms. Unfortunately, direct statistical comparisons of whole correlograms are not possible (Slatkin & Arter 1991). It is possible compute statistics for each autocorrelation on a randomization hypothesis (Cliff & Ord 1981) and to assess the overall significance of the correlogram using the Bonferroni criterion (Oden 1984). In this case, then, statistical significance of a single value of Moran's I indicates that the magnitude of the autocorrelation is greater than the mean expected under no autocorrelation ($E(I) = -1/(n-1)$, where n is the number of populations; in these experiments, $n = 25$ and $E(I) = -0.04167$) and the overall Bonferroni significance of a correlogram indicates departure from spatial randomness.

Two of the conditions simulated in these experiments, directional selection against the recessive homozygote and balanced selection against both homozygotes, should produce correlograms with *no* spatial structure – i.e., autocorrelations for all distance classes should be zero due to uniform selection (directional or balanced) on all populations. Isolation by distance generates positive autocorrelation at near distances and negative at great

MORAN I

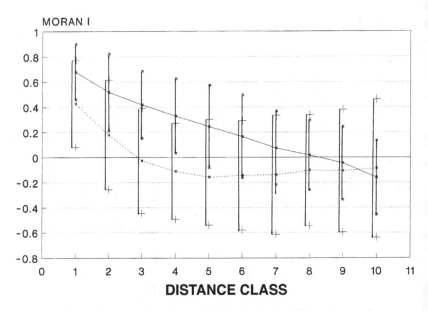

DISTANCE CLASS

Figure 4.6. Average correlograms, clinal selection (solid line) and neutral
(dashed line). (From Fix 1994, fig. 3.)

distances (Sokal & Wartenberg 1983) and clinal selection produces declining positive autocorrelations across space. Given these expectations, it
should not be surprising to find only one of the fifty balanced correlograms
to be significant at $p < 0.05$ (Bonferroni criterion). Correlograms of loci
subjected to directional selection (of lesser magnitude than the balanced
case) are significant only 14 percent (7/50) of the time. Neutral loci, subject
to isolation by distance (IBD) might be expected to show significant spatial
structure. However, under the conditions modeled in these experiments,
Bonferroni significance is achieved in slightly more than 46 percent of the
correlograms. Strong clinal selection, on the other hand, results in nearly
all correlograms being significant (49/50 or 98% at $p < 0.05$). These results
confirm expectation in both the balanced and clinal selection cases. However, strong directional selection against a recessive allele produces a
greater than expected number of significant correlograms and, conversely,
more than half of the neutral correlograms fail to attain significance.

Looking at individual autocorrelations for each distance class, the balanced case produced a few significant values of Moran's I scattered among
the 10 distance classes and 50 alleles – essentially random occurrences
among a large set of values. Both directional and neutral loci shared a
pattern of highly significant values occurring only for the first two distance

classes. The major impact of isolation by distance as modeled in these experiments is on near neighbors (see also Fix 1993). In keeping with the linear decline of spatial autocorrelation for clinal selection, numerous values of Moran's I extending to distance classes 3 and 4 are highly significant.

These statistical results show that detecting microevolutionary process from spatial patterns may be very difficult under some conditions. Patterns exist in the data but easy separation into selected or neutral alleles is not always possible. Thus, a nonsignificant correlogram may represent a neutral locus for which the dispersive forces reduce autocorrelation below significance or it could result from balanced or directional selection depressing autocorrelation.

If spatial autocorrelation is to be a useful tool in distinguishing between different classes of microevolutionary processes in real populations, it should surely perform appropriately under the artificially simplified conditions of a simulation model, particularly when the model ignores the additional statistical variation arising from sampling of individuals to obtain gene frequencies in natural populations.

In a larger sense, there may be no *general* solution to the problem of recognizing process from current gene distributions. The simulation experiments discussed here required specific numerical estimates for the various population parameters and modeled a migration process (kin-structure) that may be limited to only some societies (but see Chapter 2 and 3 for discussion of this point). The smaller the population sizes, the greater the effect of intergenerational drift. Similarly, different migration rates and patterns may increase or decrease the stochasticity of migration. Dispersive forces in populations similar to the Semai may produce considerable microdifferentiation even for relatively strongly selected loci. Hemoglobin E and ovalocytosis, two alleles conferring malarial resistance (Livingstone 1985), are quite variable among Semai settlements (Fix & Lie-Injo 1975). Similarly, Ward & Neel (1976) found statistically significant clines in the Yanomamo that they attributed to chance effects. On the other hand, larger populations with less structured migration and consequently lower levels of local genetic variation might allow better partition of effects.

Another variable to be considered along with population sizes and migration rates is the spatial scale of sample locations. The distances over which migration occurs varies greatly among human populations (see Chapter 2 for examples). This point is especially pertinent when real populations are compared to theoretical expectations (Epperson 1993). If most migration is highly localized among near neighbors within a few kilometers or dozens of kilometers, distance classes comprising several hundred kilometers would be inappropriate. Signature correlograms for isolation by distance (rapid decline of autocorrelation over the first few distance classes) have been demonstrated in simulation studies of small populations and local migration. In contrast, several recent empirical studies of spatial autocorrelation have used very large distance classes. For instance, Sokal *et al.* (1989a) in their study of European populations, set their distance classes at hundreds of kilometers. If the actual mean migration distance per generation for European populations was nearer tens of kilometers rather than hundreds, expanding the scale of distance classes to 700 km compresses all the interesting spatial genetic variation into the first distance class. Large scale migration, demic expansion, or continental clines might account for spatial patterning over such large distances but distances which are so much greater than the likely mean migration distance are entirely unrealistic for IBD.

Wave of advance of advantageous gene

The classic population genetics migration models described in Chapter 3 were based on the assumption that all loci are selectively neutral. Gene flow or migration in these models counteracts genetic drift, the dispersive force differentiating allele frequencies among groups of subpopulations. Stabilizing forces required for the systems to reach equilibrium generally have been envisioned as long range migration although it is sometimes mentioned that natural selection could also provide the systematic pressure to prevent random fixation of local groups.

Although much human genetic variation seems to be non-adaptive, considerable evidence exists for many strongly selected genetic variants including most notably those maintained by resistance to malaria. Several hemoglobin alleles (Hb S, Hb E, Hb C, and possibly others), thalassemia (several variants of both the α and β chains), G6PD deficiency, and ovalocytosis have been demonstrated to differing degrees of certitude to have selective advantage in the presence of malaria (Livingstone 1985). Similarly the Duffy blood group negative phenotype resists vivax malaria (Miller *et al.* 1975) and at least one HLA variant has been implicated in malarial resistance (Hill 1991). As Haldane (1949) pointed out long ago, the great selective potential of disease surely has affected human genetic variation. Migration models that consider only random drift would be inappropriate in these circumstances.

The classic treatment of migration–selection interaction was Fisher's (1937) paper on 'the wave of advance of an advantageous allele'. Fisher showed that migration would diffuse an adaptive allele in a constant wave of spread across space through time. His model assumed a linear, continuous, uniformly distributed population (analogous to the continuously distributed population of Wright's isolation by distance model) allowing the wave of expansion to be approximated by a differential equation such as those used in diffusion problems in physics. As he noted, in reality diffusion is often a much more complicated process. Computer simulation provides one means of introducing (and testing for the relevance of) some of this complexity. Particularly where several evolutionary forces may be interacting, simulation allows models incorporating the joint effects of this interaction. To illustrate this usage, I will consider the diffusion of the β-globin variants using the simulation models of Livingstone (e.g., 1969; 1989) and Fix (1981).

One of the many contributions of Livingstone's (1958) study of the sickle-cell hemoglobin distribution in West Africa was the realization of the importance of gene flow as the mechanism by which this adaptive allele was spread. If malaria provided the selective agent, and agriculture provided the ecological conditions for malarial transmission, then selection for the sickle cell gene commenced only relatively recently (in evolutionary time). Further, if as seemed likely, Hb S arose as a single mutation, it must have diffused throughout the range of its present distribution via gene flow within the last few hundred generations. This raised the question of the rate of diffusion of this and other adaptive alleles (Livingstone 1989). Livingstone has extensively explored this question with a series of simulations (e.g., 1969; 1976; 1989).

One of these models (Livingstone 1989) simulated selection and gene flow in a linear series of 600 populations extending from West Africa to Southeast Asia. Each artificial population comprised 500 individuals inhabiting an area 10 miles (*c*. 16 km) in diameter and thus a population density of 5 persons per square mile, similar to small-scale cultivators of the present (see Chapter 2) and thought by Livingstone to represent the demographic situation of circa 1000 AD.

In contrast to the strict linear stepping-stone model where genetic exchange occurs with nearest neighbors only, Livingstone's model featured gene flow with the three populations on either side of the reference group. As for most simulations where a number of experiments with different parameter values are carried out, Livingstone used various values for the migration rates. However, a common regime was a total migration rate of 16 percent apportioned as 5 percent with nearest neighbors, 2 percent with populations two steps away, and 1 percent with populations separated by three steps. An additional feature of this model was a random long-range migration component – migration (varying from 1 to 5%

between experiments) occurred each generation with a single population randomly chosen from those within 20 (or 30) steps of the reference population. Since several hemoglobin variants were of interest in this long transect of populations, Livingstone had to estimate fitness values for a large collection of genotypes. Fitnesses for heterozygotes ranged from 1.20 for Hb AS, 1.12 for AE, and 1.10 for AC and thalassemia heterozygotes. For homozygotes, the range was from 0 for SS to 0.85 for EE. Initial gene frequencies for the Hb alleles was 0.001 for three different populations, one in West Africa, one in the Middle East, and one in Southeast Asia representing a single mutant (one heterozygote in a population of 500) for Hb C, Hb S, and Hb E respectively. Mutation was also programmed at rates from 10^{-8} to 10^{-9} for the individual variants and 10^{-6} for thalassemia, but as Livingstone notes, except for thalassemia, mutation had little effect on the distributions.

As opposed to analytic or algebraic models, simulation models are numerical. Any particular replicate run or realization of a process must specify the values for each parameter of all processes being simulated. Especially where stochastic factors are a feature of the model, each run can be expected to be different from all others. Thus many replicates must be made to ascertain the ranges of variation in the results. Different parameter values should be explored to assess the sensitivity of outcomes to changing quantities. The results of a series of these experiments are more like a set of empirical observations rather than an equation in an analytical model. Thus the greater realism possible with a simulation model bears the cost of less generality, and the conclusions of such modeling must be more limited than algebraic results. Given these caveats, it is often possible to reach reasonably strong conclusions from simulation, particularly when different realizations show common features despite differing initial conditions or parameter values.

Thus Livingstone (1989) found that several features of the distribution of the hemoglobin variants were relatively constant despite the random elements in mutation and long-range migration in the model. The strong selection for heterozygotes for all three hemoglobin alleles guaranteed that they would increase in the populations in which they occurred. Hb S, not surprisingly due to its higher fitness, reached equilibrium in the founding population and began to diffuse before the other variants. Irrespective of migration rate, most populations began to show equilibrium frequencies for the local variant after about 120 generations (*c.* 3000 years), a not unreasonable length of time for the duration of the selective conditions produced by agricultural modifications of environments and human population densities. Figure 4.7 shows the distribution of these genes (Hb S, C,

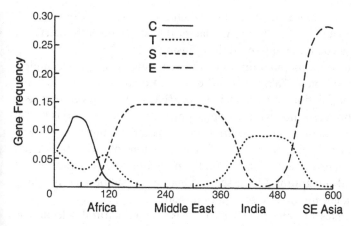

Figure 4.7. Distributions of the S, C, E, and thalassemia (T) alleles of the β-globin locus in a linear array of 600 populations after 120 generations. (From Livingstone 1989, fig. 5.)

E, and thalassemia) after 120 generations with 2 percent random migration over a range of 30 populations.

It is interesting to note that random long-distance migration rates seem to have the greatest impact on these distributions. Doubling the rate of local migration has little effect but doubling the rate of random long-distance migration greatly increases gene diffusion. Similarly, increasing the range of random migration also has a strong effect. At the highest values of long-range migration studied, a rate of 0.05 with a range of 30 populations, after 120 generations the variants have spread beyond their present geographic distributions. Earlier simulation study (Livingstone 1969) had also found that without long-range migration the spread of these alleles was quite slow. Thus the Hb E allele in these experiments took some 5000 years to move through 15 to 20 populations.

Clearly, rapid diffusion across wide geographic distances might be considerably augmented by long-distance migration leap-frogging over intervening populations to introduce the allele ahead of the wave front of advance due to local selection and migration. Unfortunately, this is a very difficult form of migration to demonstrate in extant populations. Livingstone's experiments thus provide a plausible mechanism for rapid dispersal but are hard to evaluate using empirical migration data.

Another possible mechanism to increase the rate of gene diffusion depends on the long period of time required for the adaptive alleles to increase in frequency in the populations within which they initially occur. Livingstone (1989) notes that about one-half of the 120 generations required for dispersal was spent as the variant alleles attained polymorphic frequencies in a few populations. Any factor that shortened this interval of local increase could thereby speed the diffusion of the alleles.

Definition of the population units and structure is the key to understanding this problem. Taking the entire population of Africa or Southeast Asia as the population unit, the increase in Hb C or E occurs entirely as a result of natural selection within a single population. However, this model presumes that these large populations were panmictic units, which was almost surely not the case. Many different partitions of these continental groups are conceivable. Depending on the population density and level of socio-cultural integration (discussed in Chapter 2), subpopulations may range from a few hundred to several thousand.

The dynamics of gene frequency increase under natural selection is nonlinear (Cavalli-Sforza & Bodmer 1971). Thus the time required for an allele to increase from 0.001 to 0.01 within a population is approximately equal to the time to go from 0.01 to 0.10. In a large population of 10,000, a single mutant represents a gene frequency of 0.000025. Disregarding the probability of chance loss, even the strongly selected sickle-cell gene would take a very long time to reach equilibrium. Livingstone (1976, 1989) found that to replicate Hb S diffusion in a time span commensurate with the history of malaria, he needed to begin his simulations with much higher gene frequencies. In the 1976 study, he used initial gene frequencies ranging from 0.02 to 0.005, and in the 1989 study, he began with one mutant in a population of 500 ($q = 0.001$).

The relationships between population size, areal extent of the population, and migration rate and range all influence the geography of gene diffusion. The larger the population, the longer the duration from mutation to equilibrium frequency. Small populations may experience more rapid progress toward equilibrium gene frequency but may only occupy a small segment of the geographic range. Very isolated local populations are likely to retard the diffusion as well. As Sewall Wright (e.g., 1969) stressed, the conditions for optimum evolutionary rate constitute a balance or compromise among opposed factors.

Ideally, we could discover what was the actual demic structure of these early populations through which the hemoglobin variants were spreading. Archaeology may provide some evidence for size and distribution of settlements, but rate and extent of gene flow is not easily obtained from this source. One approach is to identify a range of possible values based on comparison with extant populations and model the dynamics of gene diffusion over this range.

Recall that Livingstone (1989) chose population characteristics approximating those of the simple agrarian systems prior to the large population increases of the last 500 years. Reasoning that the population structure of forest swiddening societies such as the Semai Senoi (see Chapter 2) might provide a good model of the early agricultural populations of Southeast Asia, I (Fix 1981) carried out a number of simulation experiments to model Hb E diffusion in Southeast Asia.

From generation to generation, Semai demes comprise clusters of related settlements with population sizes of only a few hundred. These demes are not isolated and average rates of migration among them may be fairly high. Migration is often kin-structured as groups fission from established settlements. Small population size implies a shorter time to selective equilibrium within each deme as one or a few newly introduced alleles comprise a relatively larger portion of the gene pool. A single heterozygote migrant in a population of 200 is equivalent to a new gene frequency of 0.0025 (two and a half times that of Livingstone's model). More importantly, kin-structuring increases the variance of migrant gene frequencies. Rare alleles are likely to be concentrated in families. When families or kin groups migrating as units happen to bear the rare adaptive allele, the gene frequency in the recipient population may sharply increase. For instance, consider a migrant group of siblings, a common Semai occurrence (see Fix 1978 for an empirical example of such a sib group). If three or four sibs were heterozygous for the rare allele, the impact on the recipient gene pool will be three or four times that of a single random migrant. Of course, it is possible that the particular family that migrates does not possess the allele. All other things being equal, the probability that a kin group carries a particular gene is a function of the gene frequency in the population. Most kin groups will lack the rare allele. However, because of the nonlinearity of selection, those migrant kin groups that introduce a rare adaptive allele at high frequency into a new population will significantly shorten the time to selective equilibrium. The more rapid rate of increase within the population reduces the time until the allele has reached a sufficient frequency to be transmitted to the next population. KSM potentially may increase the rate of diffusion of adaptive alleles.

Livingstone's (1969) simulation of hemoglobin dynamics modeled natural selection and migration as deterministic processes – i.e., there was no random element in the selective process nor in migration. The basic equations for these processes are:

Natural selection for the heterozygote,

$$q'_i = \frac{q_i(w_2 p_i + w_3 q_i)}{q_i(w_2 p_i + w_3 q_i) + p_i(w_1 p_i + w_2 q_i)}$$

where *p* and *q* are the allele frequencies of Hb A and E respectively in the *i*th population, *q'* is the frequency of hemoglobin E after selection, and the *w*s are the fitnesses of the three genotypes.

Migration,

$$q'_i = (1 - m)q_i + 0.4mq_{i-1} + 0.4mq_{i+1} + 0.1mq_{i-2} + 0.1mq_{i+2}$$

where q'_i is the gene frequency after migration in the *i*th population, *m* is the migration rate, and the subscripted *q*s are gene frequencies in the donor populations.

These equations presume that populations are very large. For finite populations such as those modeled here, gene frequencies smaller than $1/2N$ can not exist. No migrant gene frequency can be less than one allele possessed by one of the individual migrants. Fractions of genes may not exist or be transmitted. Therefore, this basic model was modified in order to model gene flow as the movement of discrete *individuals*. Thus if *N* is 200 breeding individuals, and *m* is 0.20, 40 persons will be apportioned to the adjacent populations as 16 to the nearest neighbor on one side, 16 to that on the other side, 4 to the population two steps behind, and 4 to the one two steps ahead. As will be seen below, this discrete form of migration does have an effect on the rate of spread of alleles.

In order to model random kin-structured migration, gene frequencies were chosen for migrant groups of relatives and then randomly apportioned to neighboring populations. Several degrees of kin-structuring were tested. For most experiments, two groups were simulated: one a group of 25 related in such ways that the total number of independent genomes represented was only 75 percent of an equivalent group of unrelated individuals; the other group comprised 10 persons but with an effective number of independent genomes of 50 percent. To simulate the gene frequencies of these groups, a value was chosen from a normal approximation to a binomial distribution with expected value being the donor population gene frequency and with variance $p_iq_i/2M_e$, where M_e is the effective number of migrants (for instance, 5 in the case of the second group). Thus the gene frequencies of these kin-structured groups might randomly deviate rather widely from the donor population gene frequency. These groups were then randomly apportioned to the adjacent populations with probabilities equal to the proportions of the deterministic equation – i.e., each kin-group had a 80 percent probability of migrating to one or the other of the two immediately adjacent populations and a 20 percent probability of moving to the next but one populations on either side of the donor population.

Also since these are small populations, gene frequencies can be expected to randomly drift each generation. Genetic drift was simulated in the model in the manner described above for kin groups except that the variance was $p_i q_i / 2N_e$ with N_e being the effective population size.

The parameter values included:

(1) Fitnesses (following Livingstone (1969) were $w_1 = 1.00$, $w_2 = 1.15$, and $w_3 = 0.70$ for the three genotypes, Hb AA, AE, and EE, leading to an equilibrium at 0.25.
(2) Linear array of 100 populations each of effective size 200.
(3) Initial frequencies in the first five populations of the array were 0.05 with all other gene frequencies being 0.
(4) Migration rate of 20 percent per generation.
(5) Duration of simulation was 200 generations.

Figures 4.8, 4.9, and 4.10 compare the spread of the Hb E allele resulting from three different models. Figure 4.8 shows the wave of advance based on the deterministic migration and selection equations presented above. In the absence of a random element in the model, the wave is smooth and symmetric. The curves were drawn every 20 generations and show the frequency of the allele in each population for that generation. The final curved line at the far right of the figure represents the distribution among the 100 populations after 200 generations. Figure 4.8 indicates that in approximately 5000 years with 20 percent gene flow and balancing selection, the Hb E allele would have reached equilibrium in about one-half and would have spread to nearly three-fourths (75) of the 100 populations in the linear array.

However, if small populations are the elements of the model, the pattern of advance depicted in Figure 4.9 is the more appropriate representation. Applying the deterministic equations with the stricture that discrete genes are the units (rather than the fractional genetic currency of the continuous deterministic model) slows the rate of advance measurably. The wave front is steeper suggesting a more rapid achievement of equilibrium within populations but the leading edge of the wave is curtailed, only reaching 45 populations in the 200 generations of the simulation. Thus small population size may retard the apparent rate of advance.

Figure 4.10 shows that kin-structuring of migration produces an extension of the spread to 65 populations compared to the 45 populations of the discrete deterministic model. Although the wave of advance in both the small population models propagates at a slower rate than the infinite-size deterministic case, these models are probably better approximations of the actual demographic conditions of the real populations. The obvious jaggedness of the curves in Figure 4.10 is due to the stochastic nature of the process with both genetic drift and random kin-structured group

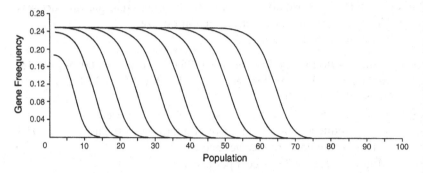

Figure 4.8. Rate of advance of hemoglobin E modeled as a deterministic process. (From Fix 1981, fig. 2.)

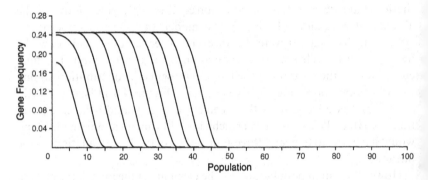

Figure 4.9. Rate of advance of hemoglobin E modeled as a discrete deterministic process. (From Fix 1981, fig. 3.)

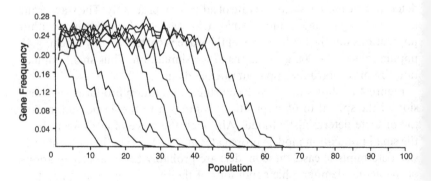

Figure 4.10. Rate of advance of hemoglobin E modeled as a stochastic process with kin-structured migration. (From Fix 1981, fig. 4.)

migration contributing to the 'noise'. Indeed, this figure has already been smoothed to some extent since the curves represent the mean values of five separate replicate runs of the program. The actual historical process in small populations would correspond to a single run and would look even messier than Figure 4.10.

These experiments reinforce Livingstone's (1969) earlier conclusion that the diffusion rate of even strongly selected alleles may be rather slow under classical near neighbor migration pattern. Kin-structuring may increase the rate but the smaller population sizes consistent with this level of analysis produce slower spread than predicted under the large population deterministic assumptions. These results pose something of a quandary since there are strong reasons for believing that malaria as a selective force was rather recent in human history. To achieve the wide distribution of several of the malarial-resistant genes in the evolutionarily short time since malaria became important required a mechanism for their rapid spread. Short-distance migration, even if kin-structured, apparently is insufficient as such a mechanism. As a consequence, Livingstone (1989) has emphasized the importance of longer-range migration to increase the diffusion rate and his simulations employing random distant migrants can replicate the distribution of the hemoglobin variants after only 120 generations.

A further factor that has been shown to increase the rate of diffusion of an advantageous allele is population growth (Fix 1981). All the previous models have assumed invariant population sizes throughout the duration of the process. However, Livingstone's (1958, also Livingstone 1962) demonstration of the importance of agriculture in fostering malaria suggests the possibility that population expansion occurred coincident with the diffusion of the adaptive alleles. Increasing population sizes and densities are thought to be consequences of the adoption of agriculture. If at the same time populations are growing, selection for malarial resistance commences, the rate of increase of the adaptive variant may be augmented. Assuming that the allele is introduced to a small population newly acquiring agriculture, the initial gene frequency may be relatively quite high allowing a rapid approach to equilibrium. Population growth might generate pressures for migration leading to the process Cavalli-Sforza *et al.* (1993) have called 'demic diffusion' (see Chapter 5 for a fuller discussion).

The effects of population growth on gene diffusion were simulated by modifying the basic kin-structured migration model. Fitnesses and pattern and rate of gene flow remained as for the previously described models. Rather than constant population size of 200 across the entire array, the initial population sizes were set at 25 persons except for the first five, which began with 100 individuals and Hb E frequencies of 0.05. These five represented 'early farming' villages in which Hb E is beginning to be selected for by malaria. The remainder of the population array were considered 'hunter–gatherers' lacking agriculture and consequently Hb E. Migrants from the 'farming' villages carried both the Hb E allele and the techniques and incentive to practice agriculture to these foraging populations, which then experienced population growth, increased malarial selection, and an increase in the frequency of the E allele.

Population growth was modeled in the simulation as a damped exponential process. Small 'foraging' populations are essentially at steady state. Growth begins with the introduction of agriculture and the Hb E allele by migrants from an established farming population. The annual rate of growth ranges from about 2 percent at the introduction of the process until becoming 0 as the population grows to equilibrium size and gene frequency ($N = 200$ and $q_E = 0.25$).

Figure 4.11 shows the effect of population growth on the spread of Hb E. The irregular wave front resulting from the stochastic effects of drift and random KSM persists, however, the extent of diffusion is much greater. After 200 generations, the allele has spread to all 100 populations and has reached high frequency in nearly 80. As for the kin-structured model (Figure 4.10), these curves are averages of five replicate runs of the program, in neither case was there great variation among runs in the number of populations reached. However, it should be noted that even under the extreme conditions of rapid population growth, limiting the migration distance to the four adjacent populations causes a slower rate of diffusion than seen in the long-distance migrant models of Livingstone (1989).

The role of simulation is to suggest plausible models for processes and to place limits on the parameter values of these processes. Further, models can focus efforts on collecting particular kinds of empirical information to evaluate them. In the case of these models of adaptive dispersal, a combination of direct data from archaeology and historical sources and better understanding of the relationships between long-distance and local migration, population growth and dynamics will help to decide among the various alternatives.

Colonization with founder effect – clines

Computer simulation offers a method to model complex population structures over time. Historical processes involving migration and/or colonization can be simulated and their effects on genetic variation can be compared to contemporary gene frequencies. Arguably one of the most important processes affecting European gene distribution was the spread

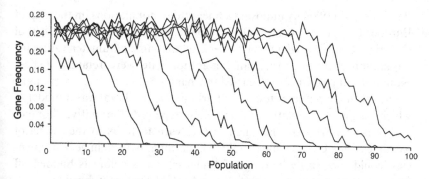

Figure 4.11. Rate of advance of hemoglobin E modeled as a stochastic process with population growth. (From Fix 1981, fig. 5.)

of agriculture from a center in the Near East (Ammerman & Cavalli-Sforza 1984; Fix 1996; Sokal & Menozzi 1982; Sokal *et al.* 1991). This topic will be extensively discussed in Chapter 5, however, since a number of investigators have used simulation to explore various models of Neolithic expansion (e.g., Barbujani *et al.* 1995; Fix 1996, 1997; Rendine *et al.* 1986), it seems appropriate to consider one of these models as an example of simulation methods in the study of migration. Not unexpectedly, I have chosen one of my own models (Fix 1997), although the results of others will be compared and discussed in Chapter 5.

One of the most salient features of synthetic maps formed from principal components analysis of European gene frequencies is a cline extending from the southeast to the northwest of the continent (Menozzi *et al.* 1978). This cline parallels the postulated spread of agriculture. Since migration is one potential cause of gene frequency gradients, it was argued that agriculture advanced by the gradual expansion of the farmers, absorbing a few indigenous foragers, but primarily through colonization of new agricultural lands in a radial wave toward the northwest (Ammerman & Cavalli-Sforza 1984). This hypothesis was formalized in an elaborate computer simulation model by Rendine *et al.* (1986) who partitioned Europe into a two-dimensional grid of populations exchanging genes in a nearest-neighbor stepping-stone lattice. From a center in the southeast, initial farming populations were programmed to grow rapidly in size and then colonize adjacent foragers' territory. After some 120 generations of simulated time, this process resulted in the replacement of hunter–gatherers by farmers all across Europe.

Barbujani and colleagues (1995) tested several hypotheses for explaining the clinal distribution in Europe. They modified the simulation model of

Rendine *et al.* (1986) by imposing a finer grid of populations on the map of Europe. Their lattice comprised 2220 populations rather than the 840 of the prior Rendine *et al.* model. Otherwise, the same parameter values for population growth, density, and migration rate were retained. (These estimates will be critically discussed in Chapter 5).

One of the hypotheses tested by Barbujani *et al.* (1995) was a model in which farming populations completely replaced the native hunter–gatherers of Europe (rather than mixing with them as in the model of Rendine *et al.*). Under these conditions, the clinal pattern in gene frequencies would necessarily be due to founder effects as colonists budded off from growing farming populations and established new farming settlements. Barbujani's group showed that this process generated spatial patterns indistinguishable from the demic expansion model and from the empirical European pattern. A potential problem with their model was the very small size of new colonies, often including only eight members. Fix (1997) argued that kin-structuring of founder groups would produce the small effective size of new populations without the problem of insufficient personnel to constitute a social group.

In order to test the effectiveness of colonization by kin-structured founder groups, I used a (by now familiar) linear stepping-stone migration model rather than the two-dimensional grid of Barbujani *et al.* The 40 populations represented a transect across Europe from the southeast to the northwest. A single population of effective size of 300 with 10 neutral alleles all with frequency 0.5 begins the simulation at one end of the linear array. This population represents the initial group of farmers whose growth and expansion will gradually colonize all of Europe.

In order to simulate colonization of new territory, population growth at the end population of the established agricultural groups produced budding of a new colony. Thus farmer populations grew according to,

$$N_t = N_{t-1}(1 + \alpha(1 - N_{t-1}/500))$$

where N_{t-1} is the population size in the previous generation, α is the growth rate, and the equilibrium population size is 500. This formula is equivalent to that of Rendine *et al.* (1986) and Barbujani *et al.* (1995) except for parameter values. Since new colonies were founded by small groups (most of the new communities of Barbujani *et al.* begin with only eight individuals; initial populations in the present study were 25), several generations of population growth were necessary before further colonization occurred. In the present experiments, when end populations reached 250 individuals, colonization of the next 'territory' in the array occurred. Population growth in the new colonies allowed growth to 250 in four generations ($\alpha = 0.777$) so that a 'wave of colonization' spread across the array at roughly one population every four generations ($= 100\,\mathrm{km}/100\,\mathrm{yrs}$ or $1\,\mathrm{km}/\mathrm{year}$, the rate of spread of agriculture found by Ammerman & Cavalli-Sforza, 1971). The colonizing groups were not a random sample of the donor population but rather a simulated kin-group. Kin-structuring reduced the effective number of independent genomes from 25 to about 12.5. This was simulated by choosing randomly the gene frequency of the colonizing group, using a normal approximation to a binomial distribution with mean, the donor population gene frequency, and standard deviation a function of the number of independent genomes (in the same manner as described in previous sections).

As colonization occurs along the linear array, the simulation sequence for each established population includes genetic drift (according to the previously described algorithm). Migration between nearest neighbors was modeled deterministically according to

$$q_i' = 0.936q_i + 0.032q_{i-1} + 0.032q_{i+1}$$

All populations smaller than 500 persons grow, and approximately every four generations, a kin-structured founding event, occurs leading to a new colony and extending the range of farmers. Since the 'indigenous' foragers do not contribute to the farmer gene pool, colonization is modeled without regard to prior inhabitants – i.e., as if empty territory was being entered.

At the end of the colonization process (approximately 160 generations), an array of 40 farmer populations have been established. Subsequent history for these populations includes continued drift and migration but no further founding colonization. One series of experiments terminated at the end of 200 generations after colonization was complete; another set was extended to 400 generations to more closely approximate the time since the process began.

The spatial pattern of gene frequencies was measured by Moran's I and arranged by distance class into a correlogram.

Figure 4.12 shows the correlograms for five replicate runs of the kin-structured founder effect simulation. Each point represents the average value of Moran's I for the 10 alleles for distance classes one to 25 for that replicate – that is, the square symbol for distance class one is the average of Moran's I for all pairs of populations in the array separated by one distance unit for each of 10 alleles in one replicate run. Similarly, the square for class two represents those populations two distance units apart for that run, and so forth. These runs were terminated after 200 generations, sufficient time for colonization to proceed to all 40 population localities in the linear array (recalling that approximately four generations of population growth were

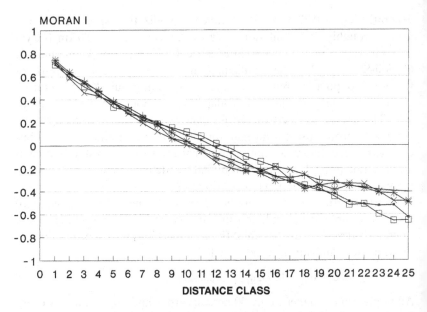

Figure 4.12. Spatial autocorrelations for kin-structured founder effect simulation model. Each point represents the average value of Moran's I for 10 alleles calculated after 200 generations of the simulation and plotted by distance class. The five lines indicate the five replicate runs of the simulation. (From Fix 1997, fig. 1.)

needed before colonization). However, it should be noted that the process of agricultural spread through Europe is thought to have taken considerably longer than 5000 years (about 200 generations). These results thus might mimic gene distributions at the end of the colonization phase rather than the present pattern of gene frequencies in Europe, which has been influenced by continuing gene flow between established farming populations.

The alleles in these runs are strongly autocorrelated at short distance and show negative autocorrelation at longer distances, exactly the pattern expected for a clinal distribution (Fix 1994, and the previous discussion on detecting natural selection). Thus, kin-structured founder effect in a linear stepping-stone population model can produce a clinal pattern similar to that generated by other evolutionary mechanisms such as demic diffusion confirming the findings of Barbujani *et al.* (1995). It should be noted, however, that the linear migration model accentuates the decline in genetic similarity with distance compared to a two-dimensional model (Kimura & Weiss 1964). Thus the extremely high spatial autocorrelations of Figure 4.12 would be weaker using a more realistic migration pattern.

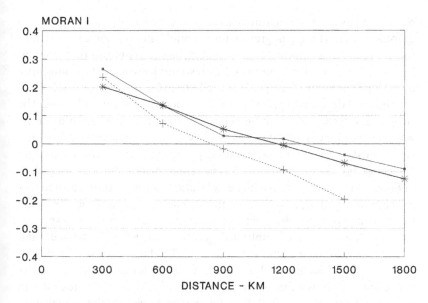

Figure 4.13. Comparison of simulated spatial autocorrelations with the empirical European gene frequencies (represented by the solid line connecting squares; drawn from data in Sokal *et al.* 1989, table 3). The dashed line connecting plus signs plots the correlogram obtained from simulation of temporal selection (Fix 1996), and the solid line connecting asterisks shows average values after 400 generations for five runs of the kin-structured founder effect model. (From Fix 1997, fig. 2.)

Figure 4.13 compares another set of runs of the kin-structured founder simulation with correlograms based on the empirical European gene frequencies presented by Sokal *et al.* (1989a). In contrast to Figure 4.12, the values of Moran's I represent the averages of all five replicates each represented by 10 alleles (i.e., each point is the average of 50 values of Moran's I for that distance class). Furthermore, these values were calculated after 400 generations of simulation, a temporal duration closer to reality. The migration rate in these runs was also increased to 10 percent per generation, a rate more in keeping with extant small-scale populations. Under these conditions, the kin-structured correlogram is nearly identical to that constructed from the empirical gene frequencies.

The fact that a simulation model can replicate reality does not mean that the mechanism being modeled actually is the cause of the distribution. Numerous models are compatible with gene frequency clines such as those observed in Europe. In Chapter 5, a number of these alternatives will be discussed. Independent corroboration is required in order to establish causation and for this case, it will be argued that evidence from archaeology as well as better comparative data from extant populations on demographic rates (including migration rates) could help to resolve the issue. In the meantime, it is sufficient to note that simulation confirms that under at least two demographic regimes and with somewhat different assumptions, founder effect can produce a clinal distribution. Clearly, the specific form of the cline will depend on parameter values chosen for the populations. The steep linear decline with distance class shown in Figure 4.12, for instance, was produced by a relatively weak flow of migration ($m = 0.032$) and represented only the foundation phase of the cline (runs terminated after 200 generations). The shallower decline represented by the simulated data in Figure 4.13 resulted from a higher migration rate ($m = 0.10$) and a longer duration of simulation (400 generations). The shorter runs included only the time necessary for the spread of farming along the entire transect but before a long period of continued genetic drift, genetic exchange and population movements after colonization. The effects on spatial autocorrelations should be maximal in the shorter runs; longer runs are a better approximation of the time since the process of agricultural spread began and it is reasonable to find that these runs better fit the actual gene frequency patterns in Europe.

Colonization with extinction – structured demes in metapopulations

Although Sewall Wright was a master of simplification in devising tractable mathematical models for evolution (Manderscheid *et al.* 1994), his conception of the evolutionary process was anything but simple. His shifting balance theory (Wright 1931) depended on the interplay between random and directed forces to create the most favorable conditions for rapid evolutionary change. This dynamic is widely accepted but some aspects of the theory remain the subject of debate. Particularly controversial is Wright's reliance on a form of group or interdemic selection as the critical mechanism causing populations to shift from one adaptive peak to another.

Following Wynne-Edwards' (1962) use of group selection to explain altruistic self-sacrifice in animal social behavior, the concept was strongly criticized by Williams (1966). Others demonstrated that the conditions necessary for group selection to predominate over individual selection were very restrictive and unlikely to occur in natural populations (Levin & Kilmer 1974; Levins 1970). Particularly important was Hamilton's (1964) elegant work that made kin selection, sharply distinguished from group selection (Maynard Smith 1976), the favored mechanism for the evolution of social behavior.

More recently, a somewhat revisionist argument has been made for widening the definition of group selection to include any 'process of genetic change which is caused by the differential extinction or proliferation of groups of organisms' (Wade 1978), an extension that would include families, kin groups, or demes as units of selection. Kinship, in this view, is a variable affecting the structuring of groups such that there is a continuum from closely related to unrelated groups of individuals.

Along with this broad definition of group selection, the concept of the population was also redefined. Levins (1970) pioneered the idea of a 'population of populations', the metapopulation, each of which is connected to the others by gene flow but are also small and subject to local extinction (see Hanski & Gilpin 1997). D.S. Wilson (1975, 1980) subdivided the population even further by considering the internal structure of breeding populations. His structured deme model (1980) is divided into 'trait groups' which are the units of selection. Trait groups may comprise a family, kin group, or 'dilute' kin group. These groups may differentially grow in size and their members thereby contribute more genes to the overall population in the next generation. The greater fitness of some individuals is not only due to their own adaptive genotypes but also is a function of the particular trait group to which they belong.

Structured deme models thus reduce the hierarchical level at which population structure is conceived. The classical models of Chapter 2 connected demes by various spatial patterns of gene flow. Trait groups are *internal* structures. Successful trait groups disperse their members to other locations ('migrate') each generation, however, not being discrete nor long-persistent breeding populations, dispersal in this case is not strictly gene flow in the classic model sense. The process, nonetheless, has consequences for the spatial dispersion of alleles.

In the previous discussion of the dynamics of the Hb E allele, various simulation results suggested that the rate of spread of this advantageous gene was rather slow under classic stepping-stone migration. Livingstone (1989) and Fix (1981) proposed different mechanisms to speed up the process, long-distance migration and KSM with population growth, respectively. Another approach to this problem made use of the structured deme concept to increase the tempo of spread of Hb E (Fix 1984). The complex population structure and interaction of evolutionary forces made this problem a good candidate for simulation modeling.

The fission–fusion pattern characteristic of populations such as the Semai Senoi (see Chapter 2) provides the basis for the model. The multiple hierarchical divisions into families, kin groups, hamlet groups, settlements, clusters of settlements, river valley groups, and so on up to and beyond the level of the ethnolinguistically defined Semai population provide many potential 'groups' as components of a structured deme model. Since much of Semai social life is influenced by kinship, many of these groups will be kin structured to some extent. This is true for both residential units (hamlets, settlements, etc.) and migrant groups that may fuse with other residential groups or may found a new such unit (lineal effect or kin-structured founder effect). Thus trait groups under these circumstances will often be dilute kin groups and group selection may grade into kin selection.

The model to be considered here (Fix 1984) depends on the division of a population (deme) into hamlet groups. These hamlets are the loci of daily life including cooperation, food-sharing, and social interaction. Among the Semai, these hamlets comprise one large or several small houses closely clumped together often inhabited by two to five related families. A common arrangement is for a senior couple to be surrounded by their offsprings' families although many other kin and non-kin groupings exist. Semai hamlets are subdivisions of settlements. The distinction is to some extent arbitrary since a very isolated large hamlet may be equivalent to a small settlement in composition and function. Generally, however, settlements are larger, from 50 to 250 inhabitants, compared to fewer than 20 inhabitants for most hamlets. The primary distinguishing feature is spatial. Most hamlets are within a short walk of others in a settlement whereas settlements are separated by a greater distance (two or more hour's walk). As previously discussed, settlements are not closed endogamous demes; because of the often close kin ties among coresidents, hamlets are likely to be exogamous. Hamlets are not good candidates for traditional population structure models since they are labile from generation to generation. On the other hand, hamlets are the basic units in ecological space and time. Hamlet members share food, cooperate in gardening and other subsistence activities, and provide each other with social support and care in illness.

This nested population structure, several exogamous hamlets contained within semi-endogamous settlements, suggests a correspondence to Wilson's (1980) model. Over the evolutionary short term, settlements may be treated as breeding populations or demes exchanging members with each other. Hamlets are the 'trait groups' among which group-specific differential viability occurs.

Many potential causes for group selection among hamlet trait groups might be envisioned. One realistic mechanism is the differential survival of group members in a disease epidemic (Fix 1984). For many diseases, whether sick individuals live or die may depend on the nursing or care they receive. A striking example of this phenomenon is the extreme mortality of Yanomamo Indians of South America (see Chapter 2) in a measles epidemic. Neel (1970; Neel *et al.* 1970) observed that the high mortality experienced by the Yanomamo was not due mainly to the virulence of the virus but to the 'collapse of village life' as most people in this measles-virgin population became ill. The basic tasks of food-procuring, -preparing, -sharing, and child-nursing and care could not be carried out. Badly nourished people were less able to survive the disease. Howell (1979:43) also explicitly notes the importance of group members among the !Kung San for survival, stating that several adults are needed in each group to 'provide insurance for families in case of adult accident or illness'.

The key requirement for invoking group selection is that survival is not simply a function of individual genotypes but is conditioned by group membership – i.e., *groups* survive or succumb. Howell's quote suggests that local group viability may be a threshold effect. For the San, if some adults are capable of providing food, water, and care for other group members, everyone may survive. Thus survival of individuals could depend on belonging to a group that included some adults resistant to infection. Those groups lacking such resistant individuals could all die due to the general break-down of social life, the 'social mortality' described by Neel. Assuming now that some individuals are *genetically* resistant to the disease agent, the differential survival and/or proliferation of these groups will increase the frequency of the resistant gene. Since several human genes are known to affect resistance to infectious disease (Salzano 1975), this is not an entirely unlikely scenario.

Still the best examples of genes conferring resistance to disease are the several hemoglobin alleles existing in polymorphic frequencies in malarious areas of the world (Livingstone 1985). One of these, Hb E, widely distributed in Southeast Asia has already been discussed and will serve as the exemplar for the model.

As for sickle-cell hemoglobin, it is the Hb AE heterozygotes that show resistance to malaria; both homozygotes are less fit, leading to a classical balanced polymorphism. In addition to the advantage that these heterozygous individuals possess, the group of which they are members has a greater likelihood of enduring an epidemic outbreak of disease. Although malaria is generally an endemic disease in the tropics, occasionally epidemics do occur (see Fix 1984 for more discussion).

For group selection to occur, genetic variation among the groups (units of selection) must be present. Even in the face of strong natural selection and relatively high migration rates, Hb E frequencies vary considerably among Semai settlements (Fix & Lie-Injo 1975). Hamlets are much more variable. Among 13 hamlets ranging in size from 6 to 35 members, the frequency of Hb E varies from 0.0 to 0.333. Since it is the hamlets that are the trait groups among which selection occurs, these highly variable frequencies allow considerable scope for group selection.

Note also that in contrast to the almost exclusive focus in the literature on altruism as the object of group selection, in the present case both individual and group selection are in the same direction. That is, heterozygous *individuals* are at an advantage in a malarious environment and *groups* with sufficient heterozygotes possess an additional advantage. Rather than the opposition between group and individual selection intrinsic to arguments about altruism, the equilibrium outcome for malarial selection of both individual and group modes will be identical. However, the *rate* of attainment of that equilibrium will be increased under the joint action of both modes of selection. (Recall that in Sewall Wright's shifting balance theory interdemic selection was seen as rapidly spreading adaptive alleles through the entire population system.) This brings the discussion back to the point raised in an earlier section: how can we account for the rapid diffusion of the hemoglobin variants across wide ranges of the Old World tropics? One of the factors affecting the rate of spread was the rate of increase of the allele within the population. Group selection augmenting individual selection can speed the attainment of equilibrium within local populations thereby shortening the dispersal time to neighboring populations.

Figure 4.14 diagrams the structured deme population model used in the simulation. The larger circles represent settlements (demes) that exchange migrants among themselves (gene flow indicated by the arrows connecting the settlements). Each settlement is comprised of four hamlets shown as smaller circles within the settlement. The hamlets are trait groups in the sense of D.S. Wilson (1980) in that they are exogamous and unstable in the long term. Each generation, trait groups dissolve as mates are found outside the groups. Hamlet members are variously related – some groups are 'family-structured' (Michod & Abugov 1980) while others are more 'dilute kin groups' (Wilson 1980). It should be noted, however, that this model differs from that of Wilson (1980) in at least two ways: settlements are treated as demes exchanging migrants; and the mechanism of group selection is hamlet extinction rather than differential proliferation.

The simulation began each generation with the 'founding' of each hamlet. This was accomplished by determining the gene frequency (of Hb E) using a random procedure mimicking genetic drift. After all hamlets in all settlements were founded, each was subjected to a chance of extinction based on its gene frequency. This process simulated group selection through differential group survival. For some of the experiments, an addi-

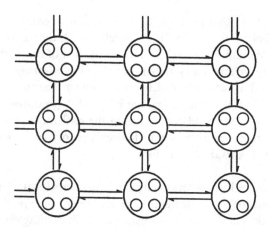

Figure 4.14. Structured deme population model. Hamlet trait groups are small circles within larger circles representing settlements. Settlements are connected by two dimensional migration between nearest neighbors. (From Fix 1984, fig. 1.)

tional step involving 'individual' selection was modeled using a deterministic equation. Finally, two-dimensional stepping-stone migration occurred between all adjacent settlements. Each of these processes will be described in more detail in the following sections.

The metapopulation of this model was an array of 25 settlement populations each of which was subdivided into four hamlets. Hamlets included 15 persons, and the settlement population size therefore was 60.

Genetic drift occurred each generation of the simulation as new hamlet gene frequencies were determined based on the settlement gene frequency of the preceding generation. Three alternative algorithms were used.

(1) RANGEN simulated founding by random individuals derived from the previous settlement gene pool with no explicit genetic relationship among them. This procedure randomly generated 15 genotypes for each hamlet in each settlement.

(2) FAMGEN modeled hamlets as sets of nuclear families (in these experiments, three unrelated nuclear families each with three children). This procedure randomly determined genotypes for two independent individuals based on the settlement population gene frequency. Using the genotypes of these 'parents', genotypes of three children were (where appropriate) randomly chosen. For each hamlet, this procedure was repeated three times to obtain the three families making up the hamlet population. Each hamlet gene pool, then, was the sum of the genes present in the families.

(3) KINGEN added further structure to the hamlet trait group by making the families of the FAMGEN model kin. Rather than three unrelated nuclear families, hamlets were comprised of three siblings, their spouses (unrelated individuals), and the nine offspring of these three families (all of whom would be first cousins). The three 'adult' siblings were first ascertained using the FAMGEN procedure. Their three spouses were randomly chosen from the settlement gene pool. From these parental genotypes, the children's genotypes were determined as in the FAMGEN procedure.

These founding procedures determined gene frequencies for the hamlets and at the same time for the settlements since settlement gene pools were simply the sum of the combined hamlet gene pools. Depending on which algorithm was utilized, random intergenerational genetic drift, or varying degrees of kin-structured founder effect was simulated.

Group selection occurred by the differential extinction of hamlet trait groups. Since epidemic disease was the selective agent and the key to group survival was having a sufficient number of resistant persons to take care of the needs of the sick, the probability of trait group extinction was made a threshold variable. For the series of experiments discussed here, the threshold value of Hb E frequency was set at 0.25, presumably assuring that several resistant heterozygotes would be available to nurse their malarial-incapacitated hamlet corresidents. Thus if the hamlet gene frequency (q_h) reached or exceeded 0.25, the threshold value (q_t), the probability of extinction (P_{ext}) of the hamlet was zero. Otherwise, P_{ext} was a function of hamlet gene frequency according to $P_{ext} = q_t - q_h$. Extinction occurred when a random number generated (from a uniform distribution between 0 and 1) was less than P_{ext}. Clearly, the higher the hamlet gene frequency, the better able they would be to cope with an epidemic and therefore the lower their probability of social collapse and group extinction. This procedure also incorporated some of the random nature of disease selection since not all hamlets with gene frequencies less than the threshold went extinct. Indeed, those groups with gene frequencies approaching the threshold would be expected to survive an epidemic (for example, hamlets where $q_h = 0.20$ would succumb only 5 percent of the time on average).

The gene pool of those hamlets that did become extinct were removed from the settlement gene pool and the settlement gene frequencies were then recalculated from the surviving hamlet gene pools. In the next generation, the extinct hamlet was replaced by a new group generated from the recalculated settlement allele frequency.

Individual selection was modeled in the manner already described in the discussion of the wave of advance of the hemoglobin genes. The fitnesses of the three hemoglobin genotypes were set to produce an balanced equilibrium at $q_E = 0.25$ following Livingstone (1969).

Migration among the settlements was simulated as a two-dimensional stepping-stone process. The 25 settlements were arranged in a 5×5 square array with migration occurring only with the nearest neighbors. Each settlement exchanged one family (or in the RANGEN series of runs, five unrelated individuals) with adjacent settlement populations. The overall rate of migration was a function of a population's location in the array. Corner populations have only two neighbors and therefore sent and received only two migrant groups, centrally located populations exchanged four groups, and laterally placed populations, three (see Figure 4.14). Migration rate varied from 0.333 for central settlements (four groups with five members each or 20 migrants from a settlement population of 60) to one-half that for corner populations. The overall average migration rate, taking into account the number of central, lateral, and corner populations was 0.267.

The genetic composition of the migrant groups was determined using the same procedure as for hamlet foundation. For RANGEN runs, five independent individual genotypes were generated based on the settlement gene frequency, and for kin-structured runs (FAMGEN and KINGEN), one family was constituted and transferred.

The full model includes several procedures (genetic drift, group selection, individual selection, and migration) as well as several alternative methods of hamlet formation and migrant group composition (RANGEN, FAMGEN, and KINGEN). This allows comparisons to be made among different sets of conditions. The primary interest is whether group selection can affect the rate of spread of an advantageous allele. However, other questions such as the role of kin-structuring in augmenting group selection may also be explored by comparing the three founding models. Similarly, the enhancing effect of individual selection may also be gauged by comparing runs with this procedure and those without. Clearly, the number of possible comparisons is very large, and those discussed here are limited to the effects of kin-structure and individual selection.

Because the model is stochastic with random components in hamlet foundation (drift), group selection, and, for the kin-structured case, migration, no single run of the simulation will be identical to another. A single realization of a set of processes represents one point in a distribution of potential outcomes. Therefore, series of replicate runs must be made to assess the degree of variation. Random decisions in the simulation model are made by comparing the probability of an event to a random number generated by a procedure in the computer. Changing the seed value from which this procedure obtains random numbers allows replication of the exact same program with a different set of random numbers, another roll of the same dice. Several such replicate runs form a *run set*. The number of repetitions will depend on the magnitude of variation among runs. In these experiments, run sets comprise five or ten replicates, depending on the comparison.

Evolution is a continuous process. The initial conditions for any model define a point in this continuum and the choice of where to begin depends to some extent on the problem of interest. Rather than begin the process at the point of mutational origin, it was assumed that all the settlement populations possessed the allele in low frequency; for these experiments the initial gene frequency was set at 0.05. Similarly, the duration of the simulation run was kept relatively short, 10 generations, since this window of time was sufficient to establish the pattern.

Six run sets were made comparing the various combinations of mode of foundation and presence or absence of individual selection: three sets *without* individual selection contrasted random group formation (RANGEN), family groups (FAMGEN), and related families (KINGEN) comprising sets F, B, L respectively; and three sets *with* individual selection added to the set of procedures and again contrasting the three foundation modes, sets G, C, K respectively.

A critical factor determining the potential for any selective process is the degree of variation among the units being selected. For group selection to be effective, there must be genetic variation among groups (Aoki 1982; Slatkin 1981). Table 4.7 shows that substantial variation (measured by the normed variance of gene frequencies, F_{ST}) and, not surprisingly, the greater the degree of kin-structuring, the greater the variance. This observation is consistent with the expectation for a greater efficacy for kin selection. Also not surprising is the much greater variation among hamlets as compared to settlements. Hamlet variation is measured among 100 hamlets, each with 15 individuals, whereas the 25 settlements are larger in size (60 persons). The ratio of hamlet to settlement averages range from a little less than twice as great (1.94 for run set G) to close to three times greater (2.82 for run set L). A further factor affecting variation is individual selection. Uniform selection across all populations would be expected to reduce variation, and comparison between run sets F, B, L (with selection) with G, C, K (no individual selection) demonstrates that selection diminishes hamlet variation (for corresponding hamlet formation type) on the order of 10–20 percent.

Table 4.7. *Comparison of genetic differentiation among hamlets and settlements among run sets*

	Variation			
	Hamlets		Settlements	
	Average	Range	Average	Range
Run sets with individual selection				
F (RANGEN)	0.057	0.040–0.075	0.026	0.015–0.037
B (FAMGEN)	0.151	0.095–0.232	0.054	0.150–0.089
L (KINGEN)	0.206	0.124–0.299	0.073	0.024–0.150
Run sets without individual selection				
G (RANGEN)	0.068	0.046–0.096	0.035	0.017–0.059
C (FAMGEN)	0.164	0.188–0.216	0.080	0.034–0.121
K (KINGEN)	0.241	0.163–0.333	0.112	0.038–0.198

The most striking effect, however, remains the mode of hamlet formation. Hamlets formed by random individuals (RANGEN) show considerably smaller values of F_{ST} than those founded by families or related families. Hamlet variation is about 2.5 times greater in FAMGEN runs compared to RANGEN runs and related family foundation, KINGEN, leads to $F_{ST}s$ some 3.5 times greater than RANGEN.

The F_{ST} values in Table 4.7 are quite high, particularly among hamlets. This variation exists despite strong individual selection in some run sets and a very high migration rate averaging 27 percent in all run sets which might ordinarily be expected to reduce overall variation to quite low levels. The extremely high variances among hamlets arise from their very small size and the process of repeated founder effect (kin-structured in two of the three cases) each generation. This great inter-group variation provides the potential for significant group selection effects.

Figures 4.15 and 4.16 show that rapid gene frequency change can be effected over the very short time period of the simulations. Figure 4.15 compares the three run sets in which hamlet extinctions occurred but not individual natural selection. Also included is the rate of increase expected within a single panmictic population subject to deterministic selection (curve labeled DET in Figure 4.15) and the curve K_{ext} which represents the single most extreme run of the ten replicates of run set K.

These results demonstrate the role of kin-structuring of hamlets on group selection. When hamlets were formed randomly (curve G), the rate of increase of q_E is slow, from 0.05 to only 0.07 after 10 generations. On the other hand, both kin-structured run sets showed much higher average

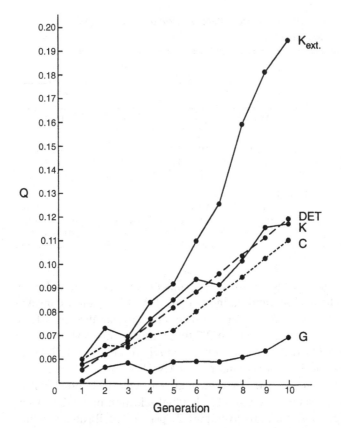

Figure 4.15. Average gene frequency (Q) increase over 10 generations for group selection without individual selection. G represents hamlet foundation by unrelated random individuals; C is nuclear family founders; K is related family founders; K_{ext} is the extreme highest single run of run set K; DET is individual selection only in a single randomly mating population (for comparison). (From Fix 1984, fig. 2.)

rates of increase. For the related family structured hamlet groups (curve K), the average rate of increase over the 25 populations was essentially equivalent to the rate for individual selection in a single randomly-mating population (DET). This outcome is consistent with the dependence of group selection on genetic differentiation – the greater variation among groups created by kin-structured founding (Table 4.7) promotes group selection.

Curves G, C, and K represent the *averages* of the 10 runs in each set and therefore provide a view of the most likely outcome of the process. However, as Stephen Jay Gould (1989) tirelessly repeats, evolution is historical,

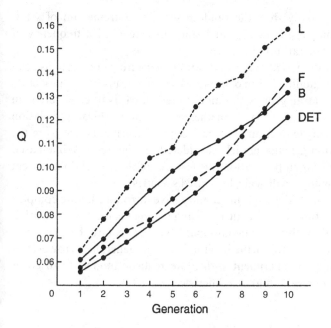

Figure 4.16. Average gene frequency (Q) increase over ten generations with both group selection and individual selection. F represents hamlet foundation by unrelated random individuals; B is nuclear family founders; L is related family founders; DET is individual selection only in a single randomly mating population (for comparison). (From Fix 1984, fig. 2.)

contingent, often dependent on chance initial events. The real world is not an average but equivalent to *one run*. From this perspective, the most extreme run of a set is not unlikely (one chance in ten) to correspond to reality. Thus K_{ext}, the extreme of run set K, suggests that evolution under group selection may be very rapid indeed as the overall frequency in this metapopulation increased from 0.05 to 0.196 in 10 generations.

Figure 4.16 shows the rates of gene frequency increase for the combined individual-group selection run sets (F, B, L). The rationale for the combined action of both modes of selection is that heterozygotes have an individual physiological resistance to malaria irrespective of their hamlet of residence while residents of particular hamlets may experience an additional 'social' cost or advantage depending on the genotypes of their coresidents. Since group selection augments individual selection in this model, the average rate of allele frequency increase for all three run sets is very steep and all are more rapid than the deterministic rate (DET). Differences due to random, family, or related family hamlet foundation are not as striking among these run sets although both the kin-structured sets increase more rapidly than the random set. The extreme run of set L (KINGEN), comparable to K_{ext} in Figure 4.15, reached a frequency of 0.237 in the 10 generations of the run.

These results show that very rapid rates of gene frequency increase can occur under this modified form of group selection. Because of the moderate kin-structuring of trait groups, this model might as easily be considered kin selection as group selection. Nonetheless, the hierarchical population structure involving extinction of some groups is clearly different from the classic population genetics models considered in Chapter 3. While based on a particular human population, the features of the model do not seem unrealistic for many small-scale, kin-based societies.

The examples considered in this chapter have introduced more complex population structures into genetic modeling through computer simulation methods. Not all problems in evolutionary genetics require these complex structures – there is great virtue in simplicity and generality. However, it does seem useful to experiment with more realistic models to discover when greater complexity makes a difference.

5 Migration and colonization in human evolution

The focus of the previous chapters has been on migration in contemporary human populations that are amenable to observation either directly by field ethnography or by written historical accounts. These studies of necessity are set in the short term and biological inference based on these observations therefore refers to microevolution rather than the longer scale of human biological history. Tribes or clusters of local groups are the standard population unit in microevolution. The effects of migration among these populations or subdivisions of a regional population are the subjects of interest. A crucial question is whether these observations on extant populations can be generalized to the long-term evolutionary history of our species.

We know that major migrations took place in the evolutionary past of our lineage that have no parallel in modern times. Initially hominids were confined to Africa and by a million or more years ago they spread to other regions of the Old World. More controversy exists about whether or not these populations were replaced by a later wave of colonizers of our own species some time after about 200 thousand years ago, but many would argue that the genetic evidence for such a displacement is very strong. More recently (and more certainly), modern humans colonized the vast empty territories of the New World and Oceania. In the cases of initial settlement, our knowledge that the event occurred is certain since range previously unoccupied by humans was filled. Only the details of these migrations (demographic and temporal) are subject to continuing controversy. As will be seen in following sections, numerous questions concerning other postulated colonizations remain unresolved.

Large scale movements within continental boundaries have also certainly taken place. Sokal (Sokal *et al.* 1996), for instance has tracked the reported movements of many peoples across Europe through history. Archaeologists have variously subscribed to migrations to explain the distribution of various cultural items including agriculture (Ammerman & Cavalli-Sforza 1971) in prehistory. Europe is not unique in this regard, evidence for mass movements in other regions of the world is extensive.

149

For instance, within the last century in Asia, peoples from South China apparently moved further south to become the 'hill tribes' of Thailand and neighboring countries. Similarly, influxes of various Indonesian peoples moved into the Malaysian Peninsula. In Africa, Bantu-speaking peoples spread to the east and south and in America Athapaskan-speakers such as the Navaho moved far to the south of their linguistic compatriots.

In contrast to these apparent major displacements of populations, studies of migration in most ethnographies record minimal movement. Most humans appear to be highly philopatric and often only short-range post-marital residence changes of one partner occur in a generation (see Chapter 2 for examples such as the Gainj). More mobile or nomadic groups such as hunter–gatherers (e.g., !Kung) may range over a wider area but seem not to undertake colonial ventures.

There seems no easy resolution to this problem. Archaeologists have grappled with the 'tyranny' of ethnographic analogy (Wobst 1978) throughout the history of the discipline, with some rejecting any use of analogy (especially for ancient populations) and others accepting varying roles for analogy (see Yellen 1977, for a discussion of the several forms that analogy may take in archaeology). The different responses of several archaeologists at the 'Man the Hunter' conference (Lee & DeVore 1968) to the revelation that the patrilocal band model did not encompass all hunter–gatherers, typify some of these contrasts. Clark Howell (1968) represents the Nietzschean extreme (if God is dead, anything goes) while others argued for better techniques to constrain interpretation including ethnoarchaeological fieldwork. Detailed ethnographic observations by persons sensitive to the problems of archaeological inference can lead to a more sophisticated application of 'analogy' or models.

Despite the difficulties in applying ethnographic analogy to past situations, it should be obvious that there is no alternative. Either a self-conscious, comparative understanding of ethnography leads to an appropriate use of analogy or implicit analogies will continue to be employed.

Explaining global human distributions requires specifying a plausible (hopefully historically accurate) migration process. At a minimum, explanations should not violate the limits of human mobility or demography. Ideally, such explanations would take into account the constraints of technology and socio-political organization as documented in comparative ethnography. The paradox is that the best data on contemporary societies do not include the large-scale migrations that seem to have occurred rarely in the past. The broader the time dimension of interest, the less these constraints may hold so that current ethnographic models may have little relevance for the events of the Pleistocene. Nevertheless, *methodological* uniformitarianism (Gould 1985) seems a necessary strategy – if *anything* goes, then all explanations are equally valid.

There is no magic solution to this quandary. As with all complex problems, multiple hypotheses and multiple lines of evidence need to be carefully evaluated to achieve explanation. In the context of studying human genetic history, understanding contemporary patterns of migration is only one component of a comprehensive strategy. Gene distributions themselves suggest historical scenarios. Linguistics, archaeology, culture history, and geography may all provide crucial information for a valid reconstruction. All of these pursuits also share a common weakness (along with the shortcomings specific to each method) in that distributions of genes, languages, or culture traits may result from many potential mechanisms. The problem of multi-causality (Weiss 1988) is always present. Although Galton's Problem (Strauss & Orans 1975) was first raised in response to Tylor's (1888) inferences about culture, the difficulty of disentangling history and function applies to all historical subjects.

Darwin succinctly stated the two causes of evolution in his phrase 'descent with modification'. Descent referred to the historical genealogical component of evolution and modification for Darwin was natural selection biasing the transmission of some characters to the next generation. Modern population genetics theory has added other components that modify the direct genealogical descent of genes including especially genetic drift. Additionally, selection may have different effects on genetic distributions depending on its mode of action. Spatially or temporally varying selection coefficients, overdominance and/or classic selection all can generate differences or similarities among populations obscuring common descent or phylogeny. As Darwin emphasized, small selective differentials may have large effects given sufficient time, however, such minimal differences in viability or fertility may be too small to detect in reasonable samples of extant populations. Thus to say that there is no evidence for selection and *therefore* a locus is neutral, is a *non sequitur*. Many 'classic' genetic loci such as the blood groups used in historical inference fall in this category. There is no *a priori* way to exclude selection and there are many tantalizing leads to selective mechanisms operating on many of these loci. Even when selection is not directly acting on a locus, close linkage may lead to hitchhiking effects. This is especially relevant to the current favorite system for reconstructing human evolution, the mitochondrial genome. Since the genome lacks recombination, it is transmitted (or not) as a single unit. Neutral variants (even those in non-coding regions) may be carried along or removed by virtue of association with selected regions.

The ubiquitous term, 'genetic marker', presumes that a particular genetic variant labels a population in some unique historical way. Just as experimental field biologists artificially mark their animal subjects and subsequently observe their distributions by noting their movements, so genes or haplotypes are treated as though they simply record the historical movement of their carriers. However true this assumption may be for specific genetic variants, it is dangerous to assume it too widely. Genes do not tell complete histories. Indeed, different genes may have sufficiently different histories to obscure *population* histories. Contrariwise, some populations may be sufficiently ephemeral over the time span of hundreds of generations to make their 'histories' meaningless.

One step toward sorting out different migration processes in human history is Weiss' (1988) distinction between local mate exchange and what he calls 'invasion'. Local mate exchange between demes, or 'gene flow' in Weiss' terminology, is a constant feature of human populations in habitats that are already settled and fully occupied. Range expansion into empty habitat or by displacing already resident populations, Weiss calls 'invasion', although in ordinary language, spread into uninhabited territory might be called 'colonization' (thus the original settling of the Americas would be a 'colonization' whereas the later occupation of these areas by Europeans constituted an invasion). Weiss goes on to relate these different forms of migration to a distinction between isolation by distance models (produced by local mate exchange, gene flow) and branching tree or 'dendrogram' analysis (caused by splitting of subgroups, expansion into new terrain, and subsequent genetic isolation). Much more needs to be said about this characterization (see below), but for now the dichotomy will be left uncritiqued. Although not corresponding to an unambiguous genetic model, Weiss also notes an intermediate pattern, that is, range expansion into occupied habitat with intermarriage and incorporation of elements of the original population into the gene pool. This is the process that Cavalli-Sforza and his colleagues (e.g., Cavalli-Sforza *et al.* 1993) call 'demic diffusion'.

To lend substance to this rather abstract argument, I will survey several current controversies in the remainder of this chapter in which large-scale, long-term migration and colonization are crucial features. I will not attempt to discuss all the dimensions of disagreement that occur in these controversies, but rather concentrate on how a better comparative understanding of migration might help resolve the issues.

Initial spread of *Homo sapiens*

In recent years more attention has been lavished on the origin of anatomically modern humans than almost any other issue in the study of human evolution. A series of conferences, edited volumes, and individual studies are products of this interest (Frayer *et al.* 1993; Mellars & Stringer 1989; Nitecki & Nitecki 1994; Stringer & Andrews 1988; Wolpoff 1989). Genetic analyses have played a major role in the debate and, indeed, are seen by some as the crucial evidence for the 'Out of Africa' hypothesis (see Relethford 1995; Rogers & Jorde 1995, for reviews of the broader genetic picture). In its extreme form, this hypothesis postulates a relatively recent origin of *Homo sapiens* in Africa followed by the invasion (in Weiss' terms – Weiss 1988) of the rest of the world and the complete replacement of indigenous populations by the invaders. At the opposite pole of the debate is the multiregional evolution model (Wolpoff *et al.* 1994) which postulates continuity between archaic and modern populations of *Homo* in the different geographic regions of the species' distribution with extensive gene flow among all populations. Between these poles, several intermediate positions have been taken including the 'weak Garden of Eden' hypothesis (Harpending *et al.* 1993) that postulates an expansion of modern humans long after the initial origin of the species.

Much of this debate will be familiar to the readers of this book and I do not propose to rehash all the evidence and arguments again. Rather I wish to focus on the specific aspects of the issue that relate to migration and colonization, in particular, the degree of isolation of early human groups, patterns of gene flow and mobility, and the roles these factors play in the different theories. More broadly, this section will continue the general discussion at the beginning of this chapter concerning the reconstruction of past human movements from genetic and other data with the focus now on the question of modern human origins.

Mitochondrial DNA

Mitochondrial DNA (mtDNA) has provided some of the strongest evidence for the African origin-replacement model. Initially from the distributions of restriction length polymorphisms (RFLPs) (Cann *et al.* 1987) and later from data on base sequences, mitochondrial variation was interpreted to locate the ancestor of all contemporary humans in Africa some 100 to 200 kya. The key claim was not only that there was a single individual female living at this time who possessed the ancestral haplotype ('Eve', the 'mother of us all' (Gibbons 1997), but also that she was a member of a *population* that was similarly ancestral to all other current human populations. From this initial population, colonizing groups spread to replace indigenous archaic forms of humans in the rest of the world.

Several lines of evidence were used to support this hypothesis (Relethford 1995), but before considering these arguments, some of the properties of mtDNA need to be specified. These special features are often claimed as advantages over nuclear DNA (and 'classical' gene loci, such as electromorphs), and while they do provide a different perspective on evolution, this gain is not entirely without disadvantages. Perhaps most importantly, transmission of the mitochondrial genome is (putatively) entirely through the maternal line and recombination between paternal and maternal chromosomes can not occur. Y chromosome DNA shares this non-recombining feature and there is considerable recent interest in tracing the evolution of markers on this chromosome (Hammer & Zegura 1996). Indeed, reference is already being made to an 'African Adam' (Gibbons 1997). This unbroken line of transmission means that theoretically one can trace all current variation back through time to a single ancestral DNA type ('Eve' for mtDNA, 'Adam' for the Y chromosome). Each variant must represent a mutational event having occurred on a framework of DNA defined by previous mutations. The sequence of such mutational events describes a branching tree, each new mutation producing a new haplotype and may serve as the subsequent frame for a later arising haplotype. These haplotype trees of contemporary mtDNA record the history of the molecule or at least that history that survived the random extinction of haplotype lines through time (Avise *et al.* 1984). It should be noted that the ability to construct a haplotype tree recording the evolutionary history of the mtDNA molecule is not identical to reconstructing the phylogeny of actual human populations (Templeton 1993). I shall return to this problem below.

The down side of non-recombination is that the entire mtDNA genome is a single linkage group and behaves in that sense as a single classical genetic locus. Thus mtDNA haplotype trees have the disadvantage of not representing the entire genome but only a part and that part subject to gender-specific events in its history (more later on this point). As previously noted in the introductory section, the fact of complete linkage of mtDNA (and Y chromosome DNA as well) opens the possibility of hitchhiking effects. That is, any variant that happened to be associated with another selected variant would be carried along or differentially removed from the pool by virtue of this association alone. This is not an unlikely scenario for mtDNA since several functionally important loci exist in the genome and several diseases result from variation at these loci (Wallace 1997). Evidence for selection on mtDNA does exist (Excoffier 1990). As will be discussed in more detail later, such selection seriously challenges the use of genes or haplotypes to 'mark' populations. Descent is faithfully recorded only by genetic variants that do not influence their own survival.

A second feature of mtDNA is the lack of the efficient repair mechanism possessed by nuclear DNA. Therefore mtDNA shows a much higher mutation rate (not shared by Y chromosome loci) than nuclear DNA, which means that considerable genetic mtDNA diversity may accumulate relatively quickly in populations (a common estimate is 2–4% change per million years). This rate has made mtDNA a useful tool for studying the early history of the species. But, like most tools, mtDNA is not an all-purpose Swiss Army knife. Despite this relatively rapid mutation rate, we shall see in subsequent sections that some recent events in human history may not be amenable to reconstruction using mtDNA – for example, the postulated spread of Neolithic farmers through Europe beginning only a few thousand years ago or the relationship between present day Southeast Asian populations.

Returning to the arguments concerning the origins of modern humans, it has been argued that the great potential diversity of mtDNA due to its high mutation rate makes it an especially useful tool for inferring modern human origins. Given the low frequency of mutation in nuclear DNA, the number of differences among contemporary humans is expected to be few, even if population divisions extend back the five million years since the split from other apes. Thus the initial findings for the Y chromosome at the ZFY locus (Dorit *et al.* 1995) showed no variation among living human groups. MtDNA variation, on the other hand, is extensive as expected on the basis of the higher mutability of the genome.

African homeland: evidence and arguments

The distribution of mtDNA variation seemed to point strongly to an African homeland for our species. Relethford (1995) identified four lines of genetic evidence to support this model. Although these patterns obtain in various degrees for other genetic systems, they are most (and sometimes only) evident in the mitochondrial data.

Relatively low genetic diversity of the human species

As Lewontin (1972) pointed out on the basis of classical loci, the amount of overall genetic variation in the human species is quite low and is mostly found within populations rather than among the major geographic subdivisions; only about six percent of the total genetic differences are among continental groups. Relethford (1995) cites more recent evidence from several sources that substantiates this figure. Limited diversity in mtDNA among human groups, then, is not unexpected. However, the extreme reduction compared to nuclear DNA or classical polymorphisms, perhaps only a third or fourth as great (Harpending *et al.* 1996), is striking. By itself, this pattern might simply suggest that a nexus of gene flow over the millennia has kept all human populations genetically similar. However, this low diversity could also result from a recent common origin of all contemporary human populations in Africa. The recentness of the events would provide insufficient time for the build-up of many mutations and founder effects resulting from the colonization of other regions would be expected to reduce gene diversity even further.

Greater genetic diversity in Africa

Supporting the recent origin model, a telling point is that the greatest mtDNA diversity is found in contemporary African populations (Harpending *et al.* 1996). On the assumption that the area of greatest diversity must be the homeland, greater African diversity signaled an African origin of the species. While the assumption that the greater the time interval since origin, the longer the time duration for differences to build up seems plausible, other mechanisms to account for genetic diversity are thinkable (Relethford 1995). Thus we return to the point made earlier that genetic distributions may have multiple causes and alternative explanations need to be carefully considered before adopting one of them.

The need to consider alternatives is accentuated in the case of greater African diversity in mtDNA since the pattern does not occur for other genetic characteristics, including blood group and enzyme loci and also the restriction length polymorphisms (RFLPs) of nuclear DNA (Harpending *et al.* 1996). Presumably a longer temporal duration since species origin should allow all genetic systems to accumulate mutations, not just mtDNA. Whether or not this observation refutes the African origin model, it does require explanation.

One possible factor is simply that many of the traditional genetic variants were first discovered in European populations. Thus these systems might show greater variation in Europe compared to Africa due to ascertainment bias – highly diverse blood group systems were not described in Africa because they were not looked for there. Rogers and Jorde (1996), however, rejected this as a complete explanation for the differences, leaving the problem of a potentially real discrepancy between the mitochondrial and the classical systems.

A second potential explanation lies in the rapid mutation rate of mtDNA. Relethford (1995, 1997) has pointed out that greater African diversity also exists in other genetic data including microsatellites, dinuc-leotide repeat loci, and short tandem repeat polymorphisms. He (and Harpending *et al.* 1996) notes that the excess heterozygosity in African populations occur for just those systems that are subject to high mutation rates. His explanation for this phenomenon depends on the relationship between population size and mutation rate in determining heterozygosity. The equilibrium formula for heterozygosity,

$$H = 4N\mu/(1 + 4N\mu)$$

where N is the effective population size and μ is the mutation rate, allows the calculation of Δ, the difference in heterozygosity between two popula-tions differing in size for genetic systems with different mutation rates (Relethford 1997). He shows that excess heterozygosity (comparing two populations of different sizes) increases with the mutation rate to a point determined by the sizes of the populations and then decreases. This sol-ution also holds in general for a non-equilibrium formulation. Thus the puzzling discrepancy between Africa and the rest of the world's popula-tions for some but not all genetic systems would be partially resolved by this argument.

In addition to mutation rates, population size differentials can also lead to different levels of genetic diversity among regions. According to the equilibrium formula above, with constant μ, heterozygosity will vary with population size. A simple explanation for greater diversity is greater popu-lation size; a larger African population over the long term would account for its increased diversity. Relethford (1995) effectively demonstrates the role of population size plotting the within-group variation for four conti-nental populations (Africa, Europe, Far East, and Australasia) against the genetic distance of each group to the mean gene frequencies over all populations (the 'centroid' of the distribution). Under the assumption of equal population sizes for all four areas, Africa shows a large positive deviation from the theoretical relationship. When the African population, however, is set at three times the size of the other groups, the relative diversities fit the prediction extremely closely.

It must also be recalled that long-term population size may be greatly affected by temporal fluctuations. The drift effective size when population size varies through time is the harmonic mean of the series (Crow & Kimura 1970:360) which is considerably smaller than the arithmetic average. Also, as Relethford (1995) mentions, if the African population grew more rapidly than other regions, its overall effective size would be greater. Again, we see here that several mechanisms might lead to the same genetic outcome.

The major problem with inferring process or history from the distribution of genes is that any one of a number of causes may produce the same pattern. In the case just considered, differences in heterozygosity depended on *both* differential mutation rates and population sizes. As Wright (e.g., 1969) has always emphasized, evolution results from the *interplay* of forces. Sorting out the specific roles of each force in the evolutionary history of a species is a complex process. Thus the superficially plausible inference that greater genetic diversity in Africa must imply a greater age for African populations and specifically an African origin for our species depends on several other things being equal.

Aoki and Shida (1996) demonstrated a critical 'other thing' in their computer simulation study of colonization and genetic diversity. Their model involved a series of successive colonizations in a linear stepping-stone of four steps representing Africa on one end and east Asia on the other. Colonization occurred when the ancestral population doubled in size and a random half moved to the next adjacent site. Migration also occurred between adjacent demes. However, as Rogers and Jorde (1995) point out, this scenario does not include a reduction in population size (bottleneck) at the time of colonization. Population sizes were equal among all demes from the point of foundation throughout the simulation. Not surprisingly, their results showed no tendency for the 'ancestral African' population to have accumulated any more genetic diversity than the descendants since all populations began with a random sample of the diversity present in the parental population. This result clearly identifies the critical requirement that colonization must be accomplished by a greatly reduced population of founders with a consequently much reduced complement of gene diversity in order for the 'diversity equals age' equation to hold. Without a severe bottleneck in population size to 'zero out' heterozygosity in the colonists, no relation between diversity and age of population should be expected.

To infer the ancestral status of African populations from current genetic diversity requires that genetic bottlenecks occurred as colonists separated to found populations in the rest of the world, a not unreasonable assumption given our understanding of human migration. However, as Rogers and Jorde (1995) show, in order to achieve a significant reduction in diversity, the bottleneck must be narrow indeed. Measuring diversity using *m*, now the mean pairwise difference in nucleotides within a population (*not* migration rate), they show that only a severe, prolonged population reduction effectively reduces diversity. Thus if the population is reduced to 33 females for a time period of 333 generations, genetic diversity is sharply curtailed. However, such extreme demographic reductions lasting such a long time seem (in Rogers and Jorde's (1995) words) 'preposterous'. They conclude that 'genetic diversity should not be interpreted as a measure of population age' (1995:24).

Thus, the argument for an African origin based on increased genetic diversity has been shown to be less compelling than it initially appeared. A larger African population combined with differences in mutation rates between some genetic systems is a plausible alternative.

Genetic distinctiveness of Africa

One of the most compelling arguments for the African origin model has been the apparent very deep root of the mtDNA tree in Africa. In the original phylogenetic tree based on RFLP analysis of 133 haplotypes, one of the two primary branches comprised only Africans and the second, which included all other types, also led to African haplotypes (Cann *et al.* 1987). Locating the origin at the base of the tree would 'minimize the number of intercontinental migrations needed to account for the geographic distribution of mtDNA types' (1987:33).

This dendrogram has been strongly criticized, particularly by Templeton (1992, 1993; see also Rogers & Jorde 1995) who pointed out that the tree was constructed without sufficient exploration of the possible alternatives. In fact, the number of these alternative trees that are consistent with the data is very large – perhaps millions, (Rogers & Jorde 1995) – and not all are rooted in Africa.

At the same time, numerous studies based on a variety of genetic materials from microsatellites to craniometrics have uncovered great diversity in Africa, leading to the conclusion that sub-Saharan Africa is the most genetically distinctive region (Relethford 1995). However, the question remains whether this distinctiveness is the result of an African origin of *Homo sapiens* or whether some other mechanism(s) are responsible. Livingstone (1991) states the problem succinctly: if genetic variation is analyzed solely by means of phylogenetic trees without regard to other forces of evolution, the interpretation of diversity will be in terms of population branching histories. Dendrograms are based on measures of genetic similarity (or dissimilarity in the case of genetic 'distances') and constitute one choice as to how to analyze differences. Given a matrix of genetic distances, a tree can be constructed. Although there are putative tests for 'treeness'

(Cavalli-Sforza & Piazza 1975), these do not necessarily exclude alternatives to the implied process of branching represented by a tree. A tree is thus a *phenetic* representation of data; the data do not contain the information that they were generated by a fissioning process. Once displayed as a dendrogram, however, the inevitable interpretation is of a historical process of splitting, isolation, and divergence. Also seemingly inevitable is the urge to locate dates of separation for the branches of the tree. Estimates of such splitting times have a long history (Weiss & Maruyama 1976) but a dubious reality.

In the absence of corroborating historical data from other sources, tree models and migration matrix models (for instance) are simply alternative ways of representing the genetic data. Any pattern of gene frequency covariances produced by a process of binary fission and divergence after isolation and thus appropriately represented in a tree form can be also be modeled by a migration matrix or isolation by distance approach (Felsenstein 1982). Identifying whether genetic differences represent one, or another, historical process (or more likely some combination) can not be accomplished uniquely from the genetic data.

The formal equivalence of the two methods has been shown in a number of studies. Livingstone (1991) simulated a set of stable populations exchanging migrants with near neighbors. With a large number of loci, he was able to construct 'phylogenetic trees' as if the genetic distances among the artificial populations had resulted from a series of binary fissions followed by isolation. Korey (1996) also simulated a migration–drift process that could be represented as very respectable looking trees. Templeton (1993:65) employed a more elaborate procedure involving nested cladistic analysis of mtDNA to conclude that the 'pattern is . . . one of restricted but non-zero gene flow throughout the cladogram'.

Templeton (1993) has summarized previous work on nuclear DNA. As already noted, many of these studies showed great diversity in Africa and dendrograms constructed from these data show Africa to be the most divergent of the major continental human populations. Templeton's conclusion is that this observed pattern is consistent with a branching phylogeny of populations but it could just as easily have been caused by long-term recurrent gene flow among populations spread across the Old World. Indeed, he states 'the only clear conclusion from the dendrograms is that gene flow has had a major impact on human evolution' (1993:67).

Reconstructing the pattern of gene flow is considerably more complex than constructing gene phylogenies. The history of human populations as a series of binary fissions is one end of a continuum of possible scenarios. The other end is defined by a network of stationary populations exchanging genes with their neighbors. Much of the empirical evidence on patterns of human migration relates to the stable end of the continuum (see Chapters 2 and 3). On the other hand, most of the recent work on molecular phylogeny, including Cavalli-Sforza and his colleagues (1994), treat history as if it was a branching process. Neither the branching nor the stable network model is likely to realistically mirror the actual history of human populations. As simplified models of reality, each can serve useful functions by suggesting hypotheses and scenarios but neither should be accepted as explanations on the basis of 'consistency' with the data.

In some cases the genetic data themselves are inconsistent with one interpretation. Bowcock *et al.* (1991:840) found that the fit of a tree model to their data on 100 DNA polymorphisms was (in their words) 'not acceptable'. While they felt that their tree model 'confirms the hypothesis that the earliest divergence in human evolution separated Africans and non-Africans' and 'a second fission . . . separates Melanesians from Chinese plus Europeans' (1991:840–1), the European data could not be so easily interpreted as a split followed by isolation and divergence. According to the assumption of their model, divergence in each separate lineage should occur at equal rates (also assuming evolution occurs by the accumulation of neutral mutations). The branch of their tree representing Europeans, however, is very short – i.e., the genetic distances between European and African populations are smaller than corresponding distances between African and all other non-African groups. Bowcock *et al.* (1991) advance two hypotheses to explain this discrepancy: (1) rates of evolution in European populations were slower (mechanism unspecified); and (2) Europeans are the result of an admixture between two ancestral populations (specifically, interbreeding occurred in one event and was immediately followed by separation). Notice that these hypotheses do not exhaust the possible causes for reduced genetic differences between the geographic region of Europe and Africa.

After dismissing the reduced rate model as untestable, Bowcock and colleagues found that the admixture model was consistent with the data – specifically, ancestral Europeans comprised 65 percent Chinese ancestors and 35 percent African ancestors. A moments reflection will verify that numerous alternative explanations for these data are possible. Clearly, the one-time mixing is a gross simplification of the likely history of these populations as Bowcock and colleagues realize. They argue, however, that more realistic models would not change the fit of their model. The question to ask, then, is whether it is the goal to 'fit the data' or to reconstruct history or to understand the evolutionary mechanisms through which historical genetic differences arose? The test of a model is only that it fit. In this case, we may ask, what does it mean to say that Europeans are a 'hybrid' population? One possibility is that Europeans are descended from some Eastern European–Western Asian populations, geographically intermediate between Africa and East Asia, and therefore engaged in a long-term nexus of gene flow through numerous intermediate populations. Fitting this model to the data with plausible numerical parameters should be possible. Assuming this fit was also acceptable, our interpretation of genetic history (I would argue) is quite different from that of the binary fission–admixture model of Bowcock *et al.* (1991). Trees are simple models and give good formal consistency with data but they also carry a very specific interpretive load: human evolution has been a branching process even if occasional episodes of 'admixture' have occurred. However, the data on human migration patterns suggest that gene flow at low but not insignificant rates is a persistent feature of our species (Lasker & Crews 1996). Unless strong evidence to the contrary exists, the choice of models for the past might better be based on present data rather than the phylogenetic models more appropriate for the evolutionary divergence after speciation.

An alternative to the bifurcation model has been provided by Relethford and Harpending (Relethford 1995; Relethford & Harpending 1994). They show that a migration model not involving population fissions can account for the existing genetic differences among human continental populations. The average number of migrants per generation required to maintain the observed genetic distances among continents was approximately 0.35 – that is, one migrant every three generations exchanged between the continental populations.

The genetic distinctiveness of African populations follows from the finding that the same *number* of migrants suffices to account for variation among each region, in other words, migration is symmetric between continents. Recall that in the discussion of within-group variation, Relethford argued that the African population must have been very much larger than those of the other regions. A larger African population size along with symmetric migration numbers means that the effective migration *rate* into Africa would be correspondingly smaller. Using the previous figure of a three-fold greater African population reduces the migration rate proportionately. This lower migration rate would be less effective in reducing the differences between African populations and the rest of the world.

Just as the model of Bowcock *et al.* (1991) discussed above was highly simplified, the migration model of Relethford and Harpending (1994) does not specify the possible complexity of the process, but Relethford (1995) notes that the numbers of migrants could fluctuate over time and space. It should also be clear that the process does not depend on frequent migration among populations at the ends of the geographic distributions. Although, as is often noted, it is highly unlikely that Inuit and San exchange mates, they are, however, part of a world-wide web of interconnected human populations (and not just electronically!).

Species population bottleneck

Negative evidence for an ancient Old World species-wide network of gene flow comes from the study of the distribution of mismatches in mtDNA (Harpending *et al.* 1993; Rogers 1995; Rogers & Jorde 1995). These studies have also shown how phylogenetic information has been obscured by the process that has generated the human pattern of mismatches. Rogers and Jorde (1995) have formulated this as an 'uncertainty principle' – the genetic data can provide information about genealogy or the history of past population sizes but not both categories simultaneously. In the specific case of human mtDNA, they interpret the pattern as the product of a major population expansion in the past. This expansion both removes the information about phylogeny from the record (removing one of the main props of the Out of Africa model) and also implies that the human population was very small at some point, too small to maintain intercontinental patterns of gene flow (thus striking a blow at the multiregional model). Having cast doubt on both models, a third hypothesis, the 'weak Garden of Eden' hypothesis (Harpending *et al.* 1993) was suggested. This scenario requires a small ancestral human population partially separated into several subpopulations living approximately 100 kya. Much later (perhaps *c.* 70 kya), these groups simultaneously greatly expanded in size. As for the other models considered here, this 'history' was inferred from the genetic data, and the various dates, population sizes, and expansions are derived from the patterns in the mismatch distributions. The primary criterion for the model, then, is the fit to the genetic data.

Figure 5.1 provides a nice illustration of the idea behind the expansion theory and, at the same time, shows how the basic data, mismatches between individual mtDNAs, are scored. The top panel (population size) of Figure 5.1 shows a hypothetical population that expanded sharply at time 7. Both population size and time include μ, the mutation rate, as a factor, since for mtDNA females are the relevant population, population size is measured as N_F; time units are $1/(2\mu)$. Taking μ to be 0.0015, population size is shown as increasing 500 fold about 58 kya. The effect of this sharp increase is seen in the second panel showing the mutational events (x) producing the gene genealogy. Most of the population, and therefore most

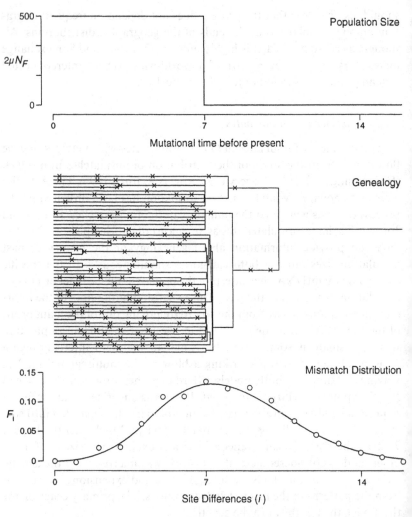

Figure 5.1. Mitochondrial genealogy and mismatch distribution of a hypothetical population. The top panel shows $2\mu N_F$ as a function of time before present, with time measured in units of $1/(2\mu)$ generations. The population was small before time 7, then increased sharply. The middle panel shows the genealogy of a sample of 50 individuals drawn from this hypothetical population (crosses are mutations). The bottom panel shows these same data as a mismatch distribution (open circles); that is, the relative frequency (F) of pairs of individuals whose mtDNA differ by i sites. The solid line represents the theoretical mismatch distribution for the parameters of the hypothetical population. (From Rogers and Jorde 1995, fig. 1.)

of the mutations, occur *after* the expansion. However, the shared ancestors are much more likely to have lived when the population was small. Thus coalescent events that are informative about phylogeny occur early in the tree and are based on only a few mutations. In the example provided in Rogers and Jorde (1995) and reproduced here, 37 of the 49 coalescent events took place before the population expansion but almost all of the mutations that characterize current mtDNA diversity occurred after the expansion. Under these circumstances, most of the information of interest (37 of the 49 events) are represented by only 7 of the total of 150 mutations. For this reason, Rogers and Jorde conclude that population growth destroys genealogical resolving power. Although this particular example is hypothetical, Rogers and Jorde contend that this is exactly the pattern of 'comb-like' or star-shaped gene genealogies that is typical (they say universal) of the empirical human mtDNA data.

The third panel of Figure 5.1 represents the distribution of mismatches among individuals. This histogram records the frequency of individuals differing at i sites where i in the example ranges from 0 to over 14. This panel also illustrates the interesting property of the mismatch distribution – the mean number of mismatched sites is equal to the time (in mutational units) of expansion. From this property, it is theoretically possible to estimate the time of expansion from the empirical mtDNA data.

To simplify analysis, the model is based on an instantaneous population expansion – i.e., the initial small population explosively grows and then remains constant in size. Rogers and Harpending (1992) realize that such demography is unrealistic but assert that the model is sufficiently robust to violations of this assumption to be useful. They also argue that major population increases are not outside the demographic potential of the human species. In particular, at an annual rate of growth of one percent (recalling that the current rate of growth of the world population is about 1.5%), a 1000-fold population increase would only take 700 years (Rogers & Jorde 1995).

Rogers (1995) performed a simulation experiment to test what parameters of a model of sudden expansion could fit the empirical mtDNA data of Cann *et al.* (1987). Using a variety of parameter values, he generated 1000 artificial data sets and found those regions of parameter space that constitute a 95 percent confidence region. Assuming nucleotide divergence rates for mtDNA of from 2 percent to 4 percent (implying mutation rates of from 7.5×10^{-4} to 1.5×10^{-3}), the interval of possible expansion times ranges from 33 to 150 kya, that is, sometime during the late Pleistocene. Prior to this expansion, acceptable parameter values for the model yield estimates of the ancestral female population of no more than a few thousand. Following expansion, the population would have included at least 150,000 females. These estimates assume that the whole human population was panmictic during this process. Geographic structure would allow for a somewhat less dramatic expansion of the population – '100-fold is enough' (Rogers & Jorde 1995). However, structured populations imply an even smaller pre-expansion population, perhaps only 1500 breeding females.

Mismatch distributions are based on comparing mtDNA pairs within a single population. It is also possible to compare the number of differences between random individuals drawn from two different populations by means of so-called *intermatch* distributions (Harpending *et al.* 1993; Rogers & Jorde 1995). Comparing the waves in the mismatch and intermatch distributions provide interesting inferences about the timing of population expansions. A pattern frequently observed in these comparisons shows a higher mean number of intermatch differences than for either within-population mismatch distribution . Figure 5.2 (Rogers & Jorde 1995) compares the mismatch and intermatch distributions of the Nuu Chah Nulth (Native Americans of the Pacific Northwest Coast) and the !Kung San of South Africa. Mismatches are represented by broken lines (dashed for the Nuu Chah Nulth and dotted for the !Kung) and the differences between individuals from the two populations (intermatch distribution) is shown by the solid line. The solid intermatch line peaks at a higher mean value and appears displaced to the right of the figure whereas both mismatch distributions show similar patterns with fewer differences.

The expansion model explains these differences between mismatch and intermatch distributions by postulating a difference between the time of subdivision of the two populations and the time of expansion of the populations. If fission and expansion occurred simultaneously, as predicted by the replacement model of human origins, the peaks in both distributions (mismatch and intermatch) should coincide. Where they differ, as for the example in Figure 5.2, Rogers and Jorde (1995) suggest two possible scenarios.

The first possible scenario is the weak Garden of Eden hypothesis, already mentioned, in which a few ancestral groups became separated but did not expand numerically or geographically until some 50,000 years after this initial subdivision. The first division on this scenario would have produced the intermatch distribution; the expansions produced the within-population, mismatch, distributions at the later point in time.

The second compatible scenario also presumes an expansion of all the groups (dated by the mismatch distributions) but does not assume an initial separation prior to expansion. On this view, the human population would have been geographically structured from the beginning of the

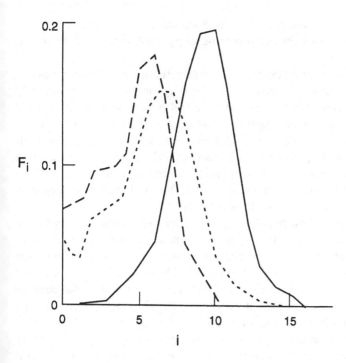

Figure 5.2. Mismatch and intermatch distribution comparing Nuu Chah Nulth (dashed line) and !Kung (dotted line); solid line shows the intermatch distribution using the same scale as in Figure 5.1. The horizontal axis represents site differences and the vertical axis relative frequencies of mismatches or intermatches. Based on mtDNA sequence data from hypervariable region 1. (Middle panel from Rogers & Jorde 1995, fig. 5.)

species with each subgroup being connected by weak gene flow. Under this model, the intermatch distributions do not date an early subdivision of the species but are the product of random introductions of migrants. MtDNA haplotypes introduced by these migrants then link them with those of the recipient group giving rise to a peak of intermatches.

A series of simulations and the confidence intervals discussed above led Rogers and Jorde (1995) to favor the weak Garden of Eden hypothesis over the geographically structured model (which bears a close resemblance to the multiregional model of Wolpoff *et al.*, 1994). The major problem, it will be remembered, with a geographically subdivided ancestral population was the extremely small population size, only about 1500 females. It seems unlikely that the wide continental distribution of humans from Africa to East Asia could have been populated by so few. Moreover, the degree of separation between the mismatch and intermatch peaks would require very weak migration among regions. Such low levels of gene flow would have generated much greater between-group genetic variation than actually exists.

As Rogers and Jorde (1995) note, however, these models do not conclusively prove (or even disprove) any of the suggested scenarios. Potentially for example, very small dispersed ancestral groups spread across wide regions may have had a pattern of inter-migration different from any modern population. The genetic patterns suggest hypotheses. At a minimum, each such hypothesis should account for the geographic distribution of gene frequencies and molecular variants. Almost invariably, more than one model can satisfy this minimal criterion. Many alternative models may coexist, each consistent with the genetic data.

In order to test these hypotheses, alternative lines of evidence need to be developed. One such source is the anthropological data on patterns of migration and population structure in contemporary groups. Too often analogies are pulled from the literature to illustrate a favored viewpoint without careful comparative study of how frequent or under what ecological and/or historical context it occurs. No single human population can serve as the model for all ancestral groups. The range of human capabilities can be demonstrated by any one human population but the most likely pattern for ancestral humans depends on a more sophisticated comparative perspective. The modern Yanomamo population may be growing at 2 percent per year (Rogers & Jorde 1995) but how likely was it for early human populations to grow over even a few hundred years at such a rate? Archaeology can possibly give us some evidence on this issue but ancient demographic patterns are notoriously difficult to document. In the meantime, conceptual models (Tooby & DeVore 1987) based on the understanding of interrelationships among many variables (population densities, land use patterns; see Chapter 2) can help us understand past migratory behavior and may provide suggestions and constraints for scenarios based on gene distributions.

Demic diffusion of agriculture through Europe

The controversy over the origin of modern humans pitted a model of population replacement (Garden of Eden) against a continuity (multiregional) model. The African origin model involved range expansion of an initially small population over the entire Old World with no intermarriage or gene flow occurring with the previous occupants (presumably a different species in any case). Gene flow played a much larger role in the multiregional model, and correspondingly, large scale movement or range expansion was de-emphasized. As discussed in the previous section, many alternative models involving combinations of colonizations and continuity are possible. One such model, demic diffusion (Cavalli-Sforza *et al.* 1993), has been widely applied to more recent events including the spread of agriculture across Europe by colonists from the Near East, the apparent expansion of Bantu-speaking populations in Africa, and perhaps colonization by agriculturalists in Southeast Asia (Bellwood 1996). These scenarios share the idea that advancing colonists do not totally replace the indigenous people but to some extent intermarry with and absorb them. Thus gene frequencies in colonized regions may be primarily those of the advancing population but also will reflect a (perhaps small) degree of genetic admixture with the prior inhabitants of the region. The prediction for the spatial pattern resulting from demic diffusion is often a cline. Gene frequencies closest to the point of origin of the spreading population will be most similar to those in the area of origin and, as distance from the origin increases, the affect of admixture (and presumably continuing gene flow with the original homeland) will cause attenuation of ancestral frequencies.

In this section I propose to examine the demic diffusion model as it has been applied to the European case, primarily by Cavalli-Sforza and his colleagues and more recently by Sokal and his colleagues (Ammerman & Cavalli-Sforza 1984; Menozzi *et al.* 1978; Sokal & Menozzi 1982; Sokal *et al.* 1991). As before, the intention is to gauge the strengths and weaknesses of the model and suggest possible alternative explanations.

The debate among geneticists over modern human origins has been focused almost entirely on the genetic data. Models have been evaluated in terms of their ability to account for the genetic patterns and history has been reconstructed from the distribution of genetic variants. It is generally noted that some interpretations of the fossil evidence fit nicely with an African origin (Stringer & Andrews 1988) but archaeology and paleontology are not equal partners in the construction of much of the theory.

The demic diffusion model

In contrast, the development of the demic diffusion model from the beginning depended on archaeological information and reconstruction (Ammerman & Cavalli-Sforza 1971). Indeed, the model stems from a classic theory in European prehistory. V.G. Childe (e.g., 1958) saw agriculture as a quantum leap in human history, a revolution that increased population density of farmers far beyond that attainable by hunter–gatherers. The complex of domesticates and techniques comprising historic European agriculture did not originate there but in the Fertile Crescent of the Near East. Due to the technical superiority provided by agriculture, Near Eastern populations burgeoned, expanded in a northwesterly direction, and ultimately replaced the indigenous foraging populations.

Ammerman and Cavalli-Sforza (1971) used carbon-14 dates on Neolithic sites in Europe to quantify the spread of agriculture. As distance from the Near Eastern origin increased, dates for initial agriculture became more recent, describing a wave of advance very like that modeled by Fisher (1937) for advantageous genes (see also Chapter 4). Such a wave might well describe the gradual diffusion of agricultural technology to neighboring populations without any movement of people at all. However, Cavalli-Sforza and his colleagues (e.g., Menozzi *et al.* 1978) followed Childe in presuming that the Neolithic was spread by colonizing farmers, initially from the Near East, later from agricultural settlements in Europe. The dynamic for this spread was also identified by Childe, the increased population densities and growth rates of farmers. Rapid demographic increase led to pressure on land, settlement fission and colonization of (by farmer standards) relatively empty forager's territories along with the assimilation of some few indigenes. This population expansion process they called *demic* diffusion in contrast to *cultural* diffusion, the spread of cultural ideas and items without any movement of people.

The demic diffusion process, then, is a perfectly respectable hypothesis for the spread of agriculture in Europe and, in fact, has old roots in archaeological thinking. As will be discussed below, however, the actual archaeological evidence for demic diffusion is not overwhelming and opinion among archaeologists on the mechanism of agricultural diffusion seems divided.

The demic diffusion hypothesis does have a clear genetic prediction: if people, in addition to crops and methods, moved, their genes must also have spread. Thus a genetic signature of population expansion should exist in the patterns of contemporary European gene frequencies. As Menozzi *et al.* (1978) discovered, there are indeed clinal patterns apparent in multivariate analyses of gene frequencies that fit the proposed southeast to northwest direction of agricultural expansion. This would seem to be strong evidence in favor of demic rather than cultural diffusion. If only wheat and barley, cows, sheep and goats, and pottery-making spread, we should not expect to see any effects on contemporary genetic variation. Thus, in this case, the model not only fits the genetic distribution, it also meshes with the archaeological evidence, and with the widely held view that farmers have a much greater population growth potential. The claimed advantage of the model is that it is compatible with the non-genetic evidence and uniquely able to explain the genetic pattern. However, those who recall the discussion in Chapter 4 of the simulation model of colonization with founder effect are already aware that alternative models for the European gene frequency clines have been developed (Barbujani *et al.* 1995; Fix 1997). These models share with the demic diffusion hypothesis a reliance on population expansion as the mechanism generating clines. It would seem that people must move in order to affect the spatial pattern of genes and therefore the cultural diffusion of agriculture would be incompatible with the genetic data. This is not necessarily so (Fix 1996).

Genetic patterns in Europe

It has been asserted that a genetic cline paralleling the direction of spread of agriculture is present in Europe. However, this pattern is not present in every genetic system. A classic textbook example of a cline is the gradual decrease in the frequency of the B allele of the ABO blood group system from Asia across Europe (Cavalli-Sforza & Bodmer 1971), but rather than following the southeast–northwest direction of the diffusion of agriculture, this cline runs east–west. In fact, *most* gene loci that have been studied in Europe do not show the southeast–northwest pattern. The data in support of the demic diffusion model were from some 10 loci with 38 alleles including the familiar blood groups such as ABO, Rh, MN, but over half (21) of the alleles were from the two histocompatibility loci, HLA-A and HLA-B. Genetic patterns were discerned not from analysis of each individual locus but from principal components analysis of the entire data set. The principal components extracted from this analysis served as 'synthetic variables' presumably representing some underlying causal factor that created the pattern. It should be noted that not only were the HLA loci preponderant numerically, they also were the only system for which there were no missing values; other loci required interpolation to fill in the lack of data points. Since it is spatial similarity that is being calculated, interpolation presents a potential source of serious bias. The HLA loci, then, are doubly important to the analysis since the complete coverage of all populations for this system should avoid this problem.

These data do demonstrate a gradient in the predicted direction. Menozzi *et al.* (1978:789) state: '[T]here is a remarkable similarity between the map of the first principal component . . . and the archaeological map of the advance of early farming'. This synthetic variable explained about one-third of the total variability present in the data and thus suggests a major factor consistent with the demic diffusion hypothesis. Remembering the importance of the HLA loci, it is interesting to note that when the total set of loci is split into an HLA component and all the rest, it is the HLA data that most strongly affect the first principal component. The non-HLA loci show a mixture of east–west (recall the B allele cline) and southeast–northwest trends in the most important first component.

Sokal and his colleagues (Sokal & Menozzi 1982; Sokal *et al.* 1991) have also studied the spatial pattern of gene frequencies in Europe. Using spatial autocorrelation analysis (see Chapter 4 for a description of this technique), they analyzed the same HLA data set as Menozzi *et al.* (1978) and, not surprisingly, demonstrated significant clinal structure. Sokal *et al.* (1991) further tested the demic diffusion hypothesis using 26 loci and examining the fit locus-by-locus. Of the 26, six loci showed spatial structure beyond simple geographic proximity and consistent with demic spread (again, two of the six loci were HLA-A and HLA-B, each with many alleles).

On the face of it, that only six of 26 loci fit the model might be taken as surprising since the classic argument for the distinctive signature of migration is that *all* loci are equally affected (this point will be further discussed below). However, Sokal *et al.* (1991) make the reasonable point that only those alleles that differ sufficiently between the immigrant farmers and the resident foragers would leave visible traces of the demic diffusion. Simulation experiments (Sokal *et al.* 1989b) showed that the initial differences in gene frequency must be greater than about 0.4 in order to be detected; this is a rather substantial difference. Thus if Mesolithic foragers did not differ greatly in allele frequencies from the Neolithic immigrants, there would be no record of the demic diffusion process. Furthermore, population movements and upheavals could obscure the pattern initially established by demic diffusion. These circumstances notwithstanding, the genetic evidence for demic diffusion is not overwhelming.

Temporal selection model

What alternative is there to demic diffusion (or the founder effect discussed in Chapter 4) to explain the spatial structure of European gene frequencies? A cline is simply a geographic gradient in a characteristic, a correlation between space and genetic similarity. Migration or gene diffusion can account for this correlation, but one of the most important causes for such patterns is natural selection (Endler 1977). One of the best known cases of clinal patterns in humans is the sickle cell hemoglobin gene resulting from environmental gradients in malarial intensity (Livingstone 1969). Interestingly, the same group that championed the demic diffusion model has proposed (in another context) that a cline in several HLA alleles was likely a result of selection (Piazza *et al.* 1980).

Since the HLA loci were major contributors to the cline shown by the first principal component, a suggestion that these genes are not simply neutral markers of population movement or expansion needs to be taken seriously. The evidence for the action of natural selection at these loci goes far beyond the Piazza *et al.* (1980) suggestion (see Klein 1986 for a review).

Several lines of evidence implicate particular HLA alleles with susceptibility or resistance to specific diseases. A recent demonstration that one allele, HLA-Bw53, provides protection against malaria (Hill 1991) is particularly noteworthy. Such demonstrations are notoriously difficult, however, since the sample sizes needed to demonstrate selection of slight magnitude are enormous. Just such small selective differentials maintained over the long term cause major evolutionary changes; indeed it was this insight that formed the basis of Darwin's theory.

The indirect evidence for the role of natural selection in shaping variation at the HLA loci is very powerful (Nei & Hughes 1991). This inference is based on the very great diversity of alleles of the HLA loci, considerably greater than would be predicted on the basis of neutral theory. Most importantly, the majority of this variability is due to substitutions in the functionally critical portion of the molecule, the binding cleft into which antigens fit (Weiss 1993). The pattern of nucleotide substitution at these antigen-recognition sites is remarkably different from that in other regions of the molecule implying natural selection as the causal mechanism (Imanishi & Gojobori 1992).

Selection for diversity in the HLA system can operate through two mechanisms: immune response over-dominance; and frequency dependent selection by pathogens (Potts & Wakeland 1990). The argument for overdominance is based on the ability of heterozygotes to respond to twice the number of antigens as homozygotes. When many different parasites confront the host, heterozygosity would be selected for. Frequency dependent selection occurs when parasites adapt to the most common host phenotype, thus selecting for those individuals with rarer alleles differing in immune response. With less-frequent alleles always at an advantage, diversity would be maintained.

The strong likelihood that the alleles providing the key evidence for demic diffusion are subject to natural selection has serious consequences for the model. If these alleles are not merely neutral 'markers' passively transported by the expansion of Neolithic peoples, then how might the cline paralleling the direction of spread of farming arise?

Gene frequency clines may also be produced by natural selection. Geographic gradients in selective conditions are a classic explanation for clines and this is exactly the explanation provided by Piazza *et al.* (1980) for the clinal patterns they documented in several HLA alleles.

For selection to be implicated in the southeast–northwest cline observed in Europe, we should expect some environmental factor to show a comparable gradient. In the case of the HLA loci, a likely candidate for a selective agent would be disease. Among contemporary human pathogens, none fit this geographic pattern. However, I (Fix 1996) have suggested a model of a *temporal* gradient in natural selection capable of producing a *geographic* gradient in gene frequencies. The argument for such time sequential selection fits nicely the diffusion of agriculture through Europe.

Menozzi *et al.* (1978) in support of their demic diffusion argument, argued that the cultural diffusion of agriculture should not have had any affect on gene frequencies '*except for selection due to the changed way of life*' (p. 786; emphasis added). The dynamic of demic diffusion depends on an extreme demographic change, massive population growth producing periodic budding, and extension of the farming population ever outward. But farming also changed the relationships of humans to the environment in many other ways. Livingstone's (1958) classic study interrelating the effects of agriculture on the ecology of mosquito vectors and consequently on selection for sickle-cell hemoglobin is probably the best known case. More recently, I (Fix 1996) argued that the spread of agriculture from the Near East to Europe changed the way of life of the people adopting it, thereby providing the selective conditions for diversity selection at the HLA loci and very likely for other classic immunological 'markers' as well.

In contrast to the rest of the world, Near Eastern and later European agriculture included both animal and plant domesticates. As this mixed farming regime spread, people and herd animals became intimately associated. Increased contact with animals increased disease transmission and many human diseases seem to be derived from domestic animals. The rise of new diseases passed from cattle, sheep, and goats to humans for which foragers had no previous experience could have changed the selective environment for a variety of human genes (see Fix 1996 for more examples).

Following this scenario, the argument that the cultural diffusion of agriculture could not have affected human gene diversity is invalid. The genetic cline does not validate the demic diffusion argument since natural selection is also capable of generating the gradient.

The hypothesis that natural selection could replicate European HLA patterns was investigated by computer simulation (Fix 1996). Using a linear stepping-stone model (see Chapter 4 for similar programs) of 168 populations (each with 300 inhabitants) representing a transect from the Near East to northern Europe, the processes of genetic drift, gene flow (with nearest neighbors at a rate of 10%), and a sequentially advancing regime of natural selection were simulated. Ten alleles for each population began with frequencies similar to those of the present-day northern European HLA-B locus. At the beginning of the simulation, the first population in the array 'adopted agriculture' and animal husbandry, and began to experience a new disease environment. Selection against the common alleles (of very low intensity – from $s = 0.005$ to 0.003) and for the rarer alleles (again varying between $s = -0.005$ to -0.003) was instituted. For each new generation of the simulation, the next population in the array was added to those experiencing selection in order to mimic the spread of agriculture and the disease environment through time. Thus the process followed the wave of advance model at a rate of one population per generation, a rate similar to the calculations by Ammerman and Cavalli-Sforza (1984) for the spread of agriculture across Europe. After 168 generations, *all* populations had become farmers and had experienced selection. However, the first population had a history of 168 generations of selection, the second, 167 generations, and so forth.

Figure 5.3 shows the results of these experiments compared to the empirical data reported by Sokal *et al.* (1989a) using Moran's I, a measure of autocorrelation (see Chapter 4). Both empirical and artificial data show the signature of a cline, a linear decline of autocorrelation with distance. The fit of the model to the data is quite striking although the simulated data show a sharper decline with greater distance classes (a better clinal pattern than the real gene data). Comparisons of the actual gene frequencies at the ends of the cline in the real European populations and the simulated data again show close correspondence (Fix 1996).

Evaluation of the models

To show that a model can fit the empirical data does not prove that history actually followed the model process. Both the demic diffusion and the temporal selection models (as well as the founder effect model – see Chapter 4) *could* have produced the cline in gene frequencies across Europe. Other data or reasoning must be marshaled to decide among these alternatives.

One long-established principle that might allow us to distinguish between natural selection and migration was stated in Chapter 4 as a preamble to discussing simulation experiments to detect various forms of natural selection. The basic idea is that natural selection should be locus-

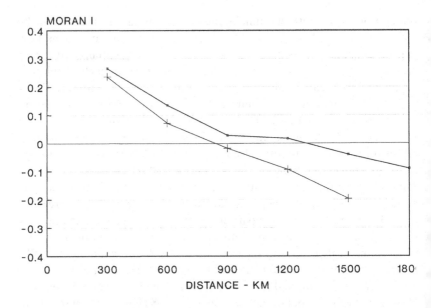

Figure 5.3. Comparison of simulated spatial autocorrelations with the empirical European gene frequencies. The solid line connecting squares represents the empirical values of Moran's I based on European gene frequencies redrawn from Sokal *et al.* (1989, fig. 2b); Morans's I was calculated directly from the data on 32 significant loci listed by Sokal *et al.* (1989, table 3) and plotted by 300 km distance clases. The dashed line connecting plus signs plots the correlogram obtained from simulation of temporal selection. (Redrawn from Fix 1997.)

specific while migration and/or genetic drift depend on population proper-ties (size and migration rate) affecting all loci equally.

In fact, from this reasoning Barbujani *et al.* (1995) reject the selection explanation. While they grant that parasite-driven frequency dependent selection could be responsible for clines, in the specific case of Europe they conclude that 'the nearly parallel gradients observed in *many* independent alleles suggest that an evolutionary pressure affecting the entire genome and not merely part of it, i.e., gene flow, played a major evolutionary role' (Barbujani *et al.* 1995: 127; emphasis added).

Surely this is a strong argument. On the other hand, 'many' is a relative concept. As I pointed out, surprisingly *few* loci actually show the trend, although there are legitimate reasons why this might be true even if migration were the true cause of the clines (mainly the problem of sufficient genetic difference between the migrants and the sedentes to be detected). Still, the argument that many loci show the trend must be tempered by the realization that only a few are generating most of the signal.

Contrary to the adage that selection should not affect many loci, natural selection of the sort proposed in this model *would* have a similar effect on many loci. The fact that the European clines for several genes are nearly parallel (Barbujani *et al.* 1995) is expected since the postulated mechanism – selection due to changed disease environments under animal domestication – diffused along the same gradient. Note also that each locus need not be affected in exactly the same way by selection, only that the frequencies at these loci be changed after domestication. Alleles at many loci might be experiencing frequency dependent selection for the same reason postulated for the HLA loci. As Haldane (1949) pointed out long ago, the changes in human life associated with the adoption of settled agriculture and particularly the increased importance of disease, increased enormously the potential for natural selection. Although it is extremely difficult to confirm small selective differences statistically, there is certainly evidence for disease selection at several of the traditional genetic 'marker' loci. For instance, the statistical association of ABO phenotypes with various stomach ailments (Cavalli-Sforza & Bodmer 1971) has now received support in the demonstration that *Helicobacter pylori* (implicated in gastritis, gastric and duodenal ulcer, and gastric adenocarcinoma) binds to Le^b antigen (Boren *et al.* 1993). Similarly Duffy negative phenotypes lack receptors for vivax malaria parasites (Miller *et al.* 1975). *Many* alleles are selected by malaria (Livingstone 1985) and can be expected to show parallel gradients. Thus a major change in disease environment potentially increasing the incidences of many diseases may have had selective effects on several loci.

In contrast to the clinal distribution of several immunological systems, recent data on mtDNA from European populations do not show comparable gradients in distribution. Indeed, the conclusion of one study was that the 'major extant lineages throughout Europe predate the Neolithic expansion and that the spread of agriculture was a substantially indigenous development accompanied by only a relatively minor component of contemporary Middle Eastern agriculturalists' (Richards *et al.* 1996:185).

Although numerous explanations are possible for the discrepancy between the mtDNA evidence and the immunological genetic data, different selective regimes is a likely cause. Among the alternatives proposed by Wise *et al.* (1997:714) to account for the low ratio of human mtDNA diversity to nuclear heterozygosity is a contrast between 'diversifying selection acting to increase variation in the human nuclear genome' and 'directional selection acting to reduce variation in the human mitochondrial genome'. This logic when applied to the European case would have frequency dependent disease selection increasing the diversity of HLA alleles and the relative differences in frequencies in other loci between foraging and farming populations. In contrast, mtDNA was not affected by the new diseases (selectively neutral) and/or, behaving as a single locus, it was directionally selected for by some other agent(s). In any case, the mtDNA data do not support the demic diffusion hypothesis.

Although not strictly genetic data, information on skeletal samples gives a direct look at the transition and should provide another measure of biological continuity or discontinuity between Mesolithic and Neolithic peoples. Unfortunately, as Jackes *et al.* (1997b) point out, only a few studies have examined this question, and these have not reached the same conclusions. Harding *et al.* (1990) found that cranial measurements did not support the demic diffusion hypothesis. Lalueza Fox (Lalueza Fox *et al.* 1996), on the other hand, stated that craniometric data do support Neolithic population replacement in the Iberian Peninsula. Jackes and colleagues (1997a), however, argue that a more complete analysis of data from Portugal does not show a sharp separation between Mesolithic and Neolithic crania. In fact, analysis of non-metric dental morphological traits 'indicates genetic continuity between the Portuguese Mesolithic and Neolithic' (Jackes *et al.* 1997a: 647). Thus the skeletal evidence for demic diffusion is weak or negative.

As already noted, the direct evidence from archaeology for replacement or continuity is likewise mixed. The agricultural complex including the domesticates such as sheep and goats, wheat and barley (not natives to Europe) as well as certain techniques such as pottery must have diffused. The actual data recovered by archaeologists include material remains of these items and the dates at which they appear in various localities. But the spread of such materials does not directly inform us as to whether they were carried by new invaders or were passed along and adopted by local populations.

Ammerman and Cavalli-Sforza (1984) have provided criteria for inferring demic diffusion. The model predicts both a marked population increase at the transition and relatively homogeneous regional cultures of farmers. Cultural diffusion would seem not to produce a population increase and we should see continuity in material culture and settlement patterns across the transition with regional differences being retained. A moment's reflection, however, suggests that these are not very reliable discriminators. The cultural diffusion of agricultural domesticates and techniques could easily induce population growth. Moreover, the diffusion of technology might be rapid and lead to the sharing of similar cultural items among formerly dissimilar foragers. Thus there are no diagnostic tests that allow us to decide between the hypotheses.

While this is not the place to review the very extensive archaeological literature on this phase of European prehistory, some indication of the complexity and lack of unanimity among scholars apparent in recent studies will be noted. Regional studies of the early Neolithic present a mosaic of different pictures of events. In some areas such as southeastern and central Europe, farming appears as intrusive and these areas may have been colonized by agriculturalists (Dennell 1992). In southwest Germany, long periods of symbiosis between farmers and foragers seem to have obtained (Gregg 1988), and in other regions the evidence points to local foragers adopting agriculture (Dennell 1992). Barker (1985) also emphasizes the continuity from the Mesolithic to the Neolithic in the Alpine area. (See Fix 1996 for further examples.) Speaking of Portugal, but extending the argument to other areas in western Europe, Jackes *et al.* (1997b: 644) state that the transition to the Neolithic 'involved a slow and piecemeal intensification of many factors already present'. Finally, Barker (1985) in his study of prehistoric farming in Europe, concluded that local Mesolithic populations played a much greater role than colonizers. Part of this conclusion was based on the uniformly low population densities of early farmers, not substantially greater than that of Mesolithic foragers. If, as these data imply, there was no tendency for rapid growth of early agricultural populations, the primary causal mechanism for demic diffusion disappears.

Perhaps the most critical assumption of the demic diffusion model is the requirement that farmers experience rapid population growth to drive the colonization process. The simulation model (Rendine *et al.* 1986) that provides the clearest perspective on the demic diffusion process depends on a nearly 30-fold increase in population size between foragers and farmers (from forager populations of 300 to farmer populations of 8000) with a corresponding increase in population density. These parameter values were derived from ethnographic analogies (rather than archaeological estimates) and forager values are based mainly on Australian aboriginal (Tindale 1953) and African Pygmy (Cavalli-Sforza 1986) data; the source for the size and densities of early farmers is less clear. The population growth rate for farmers is initially very high, becoming damped as the equilibrium carrying capacity is reached according to the formula

$$N_t = N_{t-1}[1 + \alpha(1 - N_{t-1}/8000)]$$

where N_{t-1} is the population size in the previous generation, α is the growth rate (0.5 in the experiments of Rendine *et al.* 1986), and the equilibrium population size is 8000.

These differentials between hunter–gatherers and agriculturalists are substantial. To provide some numerical appreciation of the magnitude of growth, with an α of 0.5, in one generation, population size would increase from 300 to 444 persons, an annual rate of increase of 2.7 percent. While this rate of growth is not unknown or impossible for humans, it is more typical of modern economically emerging nations rather than that of populations prior to the existence of modern public health and medicine. For example, the Semai population of farmers described in Chapter 2 experienced a much lower rate (0.7%) in the recent past (Fix 1977). The extremely high growth rates of population such as Tristan da Cunha colonizing a new habitat are not sustained over the longer term; the periodic crashes of the Tristan population, for example, are well known (Roberts 1968).

While it is likely that agriculture made possible the great increase in the human population over the last 10,000 years, it is less certain that the 'revolution' made such an immediate impact on population. If it did, the replacement of foragers by farmers (along with the absorption of a few foragers) in somewhat over 100 generations was inevitable.

Although the exact parameter values for population sizes and densities are difficult to estimate archaeologically, it is possible to evaluate comparatively the population characteristics of settlements of Mesolithic and early Neolithic Europeans. Dennell (1992:90) for instance, attacks the critical assumption that early Neolithic farmers had 'overwhelming demographic and economic superiority over local Mesolithic hunter–foragers and thus could appropriate their lands and dispossess them'. Indeed, he argues that the process of agricultural diffusion through Europe suggests a 'slow and very gradual assimilation' of agriculture by indigenous foragers rather than a swamping by waves of farmer/invaders. Similarly, Meiklejohn *et al.* (1984) estimate Mesolithic population densities in the same range as for simple farmers and argue that the demographic contrast between Mesolithic and Neolithic Europe was not sharp. Jackes *et al.* (1997b) also downplay this distinction noting the evidence for relatively sedentary Mesolithic populations (in contrast to the Australian and African hunter–gatherers used as primary sources for the Rendine *et al.* model) as well as that for the seasonal movement for foraged resources among some Neolithic populations (similar to some modern farming groups such as the Semai Senoi – see Chapter 2).

We have now seen that three hypotheses are available to account for the apparent genetic cline in several genetic loci across Europe: demic diffusion; temporal selection; and founder effect (developed in Chapter 4). All three depend on the diffusion of agriculture from the Near East, although the selection model is based on cultural transmission whereas in the other two models agriculture spreads by human movement. The distinction between the colonization models involves the degree to which advancing farmers incorporate the previous occupants – a small amount of admixture is a feature of demic diffusion, while the founder effect model requires complete replacement of the indigenes.

As for the modern human origins controversy, numerous intermediate models involving mixes and/or interactions among these mechanisms are possible. Indeed, the archaeological data indicate considerable regional variation in the pattern of population settlement and interaction suggesting a range of degrees of population replacement from complete to none. As Barbujani *et al.* (1995) and I (Fix 1996) have shown, all three models are able to explain the clinal distribution. More complex models generated by Barbujani *et al.*, involving later movements of people into Europe, did not improve the fit of the simulated to the empirical data, although archaeological and ethnohistorical evidence shows numerous such population displacements after the Neolithic. Apparently these movements were insufficient to 'erase' the signature left by the diffusion of agriculture.

Resolution of the issue depends on integrating evidence from many sources. Additional genetic data from systems other than the classical 'markers' such as HLA and the blood groups for which there is evidence of selection might give a better historical record of population movements. However, initial findings from newer molecular genetic technologies have not solved the problem. The mtDNA data, for instance, record no cline at all but this has not been accepted as disproof of the demic diffusion hypothesis. This is not to say that further molecular data might not decide the issue, but given the history of the modern human origins debate, immediate solutions seem unlikely.

Archaeology provided what seemed like a strong foundation for the demic diffusion idea since it simply formalized a classical archaeological hypothesis. In recent years, however, the hypothesis has not been universally accepted by archaeologists and some have explicitly criticized the Ammerman–Cavalli-Sforza model (e.g., Dennell 1992). Because the models differ primarily in the relationships of foragers and farmers, the recent great interest in this area by archaeologists can be expected to provide important insights.

A key issue is the demography of transitional populations. Archaeology can provide some information on this topic and the comparative ethnographic perspective advocated in Chapter 2 of this book might also help. Are Mesolithic Europeans more likely to be similar to African Pygmies/ Australian Aborigines or to the densely settled hunting–gatherering–fishing populations of the Northwest Coast of Native North America? Were early European farmers like the relatively sparsely settled, not rapidly growing, Semai swiddeners or like the annihilating wave of European colonists of the Age of Expansion displacing the indigenes (Crosby 1986)? Surely the early Europeans were not exactly equivalent to any contemporary population. Well-validated conceptual models relating population density and political integration to mobility would provide better sources for genetic models than the 'grab-bag' approach of choosing a referential analog. Thus once again we are in a similar situation to the debate on the origins of modern humans.

Prehistoric Southeast Asian dispersals: Peninsular Malaysia

The two poles of a continuum of explanations in both the controversy over modern human origins and the demic diffusion debate were *in situ* evolution and mass migration. This dichotomy persists throughout historical biology. Populations may show similarities because of common ancestry or similar environmental selection pressures. Likewise, differences may be due to migration of one group from elsewhere or local selective or random differentiation.

The explanatory importance of migration has waxed and waned in both biological and socio-cultural anthropology with 'functional' schools succeeding 'historical' schools only to be challenged again by proponents of history. Presently historical explanations depending on population movements have returned to prominence in anthropology, as illustrated in archaeology by Gamble's (1994) book, *Timewalkers: The Prehistory of Global Colonization*, and in genetical anthropology by Cavalli-Sforza *et al.*'s (1994) major work *The History and Geography of Human Genes*. These arguments have a long history and, as we have seen in the case of the diffusion of European agriculture, migrational explanations were often early favorites.

The human population of Southeast Asia presents a complex mosaic of biological, linguistic, and cultural variation. Just as for Europe, the history of human movement within and from Asia has lead to wide dispersals of peoples, languages, and cultures (Akazawa & Szathmary 1996). The human genetic diversity of Asia also was traditionally explained by waves of migration, and variants of the demic diffusion model have been extended to this part of the world as well (Bellwood 1996). The human biology of Asia, however, is not only about the migration and mixing of gene pools, local adaptations have also shaped the genetic structure of the region (Fix 1995).

According to the Garden of Eden (weak or strong) model, pre-modern Asian populations must have been replaced by moderns migrating ultimately from Africa. Representations of this process sometimes show several waves of migration in the early history of Asian moderns (Lahr 1996), a more complex scenario than the straightforward replacement of Neanderthals by Cro-Magnons in Europe. But long before the historical reconstructions based on mtDNA became possible, human variation in Asia was thought to represent waves of successive migrations. Early migrations left remnant populations in refugia into which they had been pushed by more numerous, perhaps technologically superior, later migrants. Waves were named for their putative destination (Australoids) and/or their presumed origin (Veddoid) (Cole 1945). The human populations of the region thus could be likened to a layer cake, early strata overlaid by later or, following the usage of one German historical school, successive circles of culture traits (*Kulturkriese*) like ripples on a pool (Schmidt 1939).

Although called 'historical' theories, most of these explanations were reconstructions based on the distribution of contemporary traits, biological or cultural. Instead of DNA markers, somatic or morphological traits identified the 'racial' composition of present-day populations, descendants of the past migrational movements into the region. Often these hypotheses mixed 'racial' and cultural traits together in the taxonomy; 'primitive' technologies were a marker of earlier waves.

While a linguistically remnant population, the non-Indo-European speaking Basque (see Chapter 2), has survived in Europe, no hunter–gatherers persisted into modern times. In Southeast Asia, an extreme range of technologies from foragers through almost every conceivable degree of intensity of agricultural land use from swiddeners to terraced and irrigated padi rice cultivation have coexisted. Perhaps this diversity is due to the heterogeneity of the Southeast Asian environment, hill tropical forest grading to coastal plain, providing numerous niches for hunter–collectors, traders, farmers, and fisherfolk. But whatever its cause, this complexity refutes a simple demic expansion of agricultural colonists replacing foragers.

The pattern of biological and cultural diversity present in Peninsular Malaysia is a microcosm of the broader Southeast Asian picture. Part of the mainland, but projecting into the sea world, the peninsula is connected to the history of both. Linguistically the Malayan aboriginal ('Orang Asli' or 'original people') languages are a component of Mon-Khmer, Austroasiatic languages of the mainland, while the Malays speak an Austronesian language linking them with the islands (indeed, from Madagascar to remote Oceania). This mosaic of languages, cultures, and biological characteristics provides an appropriate case to compare and contrast the roles of colonization, migration, and other evolutionary forces influencing gene geography.

The present-day majority populations of the peninsula are recent migrants to the area. The large population of Chinese descent and considerably smaller component from India came mainly in the nineteenth century in response to economic opportunities in tin mining and plantation work. Malays entered the peninsula over a much longer time period and from several different regions of Indonesia. Only a tiny minority in numerical population, the Malayan Aborigines (hereafter, 'Orang Asli') are traditionally considered to be distinct from the Malays proper. The Orang Asli are internally divided into three components: Semang; Senoi; and Melayu Asli (Fix 1995). Just as for the majority populations, these three groups of Orang Asli have traditionally been thought to represent separate sequential waves of migration to the peninsula.

The Semang are described in the early anthropological literature (summarized in Carey 1976) as one of several populations of small-statured, frizzy-haired Negrito peoples (also including the Andaman Islanders, the Aeta of the Philippines, and possibly some Papuans). These widely scattered groups were considered remnants of an ancient, once pan-Southeast Asian population now confined to small refuge areas in the forests or, in the case of the Andamaners, on islands. Thousands of years after the Semang Negritos, a second colonizing wave was said to have arrived in the peninsula. These were the ancestral Senoi peoples, of whom the Semai, described in Chapter 2, are the largest current component. For a number of reasons, Senoi were thought to be related to the Veddas of Sri Lanka and/or Australian Aborigines (Skeat & Blagden 1906). Some 4000 years ago, proto-Malays (Melayu Asli) were said to have displaced earlier Senoi and Semang groups from parts of the peninsula, only to be swamped themselves within the last few hundred years by so-called 'deutero-Malay' colonists from Indonesia.

Each wave of colonists was seen as a distinct cultural and biological unit with biological markers carried from the original homeland and with a technology and culture appropriate to the presumed antiquity of entry into the peninsula. Semang, as members of the original Negrito population of Southeast Asia, were short, dark-skinned, 'frizzy-haired', nomadic hunter–gatherers. The Senoi, spreading later in time, practiced simple agriculture. Since Martin (1905), the Senoi who are lighter in skin color and 'wavy-haired', were considered to be distinct racially from Semang and Malays. The Melayu Asli were thought to be even lighter in skin color and straight-haired, practicing a more intensive form of agriculture (including padi rice cultivation) and engaging extensively in trade. Thus, the traditional schema explained differences among Orang Asli groups as the result of a temporal stratification, each new colonizing wave pushing members of previous subgroupings into less desirable territory. In case these views might be considered archaic and unworthy of contemporary consideration: 'Clearly, there is, in Southeast Asia and Oceania, a dark-skinned genetic substrate, probably resulting from earlier immigrations. Some of the immigrants, living in marginal environments, remained relatively unmixed, especially where later migrations (e.g., of Malayo-Polynesians) have had a lesser genetic impact.' (Cavalli-Sforza *et al.* 1994:237).

As always, alternative explanations for this diversity are possible and need to be considered. One such alternative has been the reconstruction of the linguistic and cultural history of the peninsula provided by Geoffrey Benjamin (1976, 1985, 1986). Benjamin points out that all three of the Orang Asli groups speak languages of the Mon-Khmer group of Austroasiatic. Analysis of these languages suggests that all three groups had a common ancestor between 5000 and 6600 years ago. On this basis, he argues that these groups differentiated *in situ* from a common linguistic and cultural matrix. Based on a generalized Southeast Asian foraging tradition which the Semang (to a greater or lesser degree) continue, the other traditions took up a more settled farming life (Senoi) perhaps as much as 5000 or 6000 years ago, and in the case of the Melayu Asli, some of the more southerly and coastal populations became involved in the great Asian trade system beginning as much as 5000 years ago. According to this formulation, then, Orang Asli linguistic and cultural differences are not the result of outside migrants introducing new patterns to the region but are rather a local development as the different traditions specialized in different economic pursuits. The biological corollary of this model is that present-day morphological and genetic diversity has evolved within the region within the last 200 to 300 generations.

Again, the polar hypotheses of waves of migration versus *in situ* differentiation allow a wide spectrum of intermediate positions. As the previous discussion concluded, it is likely that both colonization and evolutionary change affected the European gene pool. Similarly, it would not be surprising if both processes were at work in Malaysia. Bellwood (1993) has, in fact, proposed that both regional continuity and 'successive population flows into Peninsular Malaysia' have occurred.

A complete examination of all data relevant to the various possibilities would require more space than is available here. However, I propose to examine in some detail the implications of Benjamin's model for the biological history of the peninsula and, more generally, the relative roles of migration and local evolution as explanations (see also Fix 1995). It should be made clear that I am not attempting to reconstruct Malayan history from biological traits, genetic or morphological, a task for which the biological data are simply insufficient. Cavalli-Sforza *et al.* (1994:238) also conclude that the genetic data for Southeast Asia as a whole are too few for detailed analysis of population histories (although they do produce a dendrogram of these populations – see Figure 5.7). The aim of this section will be to examine in so far as the data allow the plausibility of the hypothesis of local evolution and to assess the role of the socio-cultural features on the biology of this multi-population system. This latter goal has implications for the broader problem of studying and inferring isolation and gene flow over long stretches of human history.

Cultural traditions in Peninsular Malaysia

Benjamin's (1986) reconstruction of the past social patterns of the peninsula (its 'paleosociology') rests fundamentally on the linguistic history of these subgroupings of Austroasiatic languages. Assuming all three presently recognized categories – Semang, Senoi, and Melayu Asli – share a common linguistic ancestor, it follows that they must also share a common cultural pattern (and perhaps biology). The assumption of a single speech community ancestral to a group of current related languages is widely accepted in linguistics and, for example, is the basis for the search for an Indo-European homeland (Mallory 1989).

Benjamin (1986) theorizes that the adoption of differing subsistence modes was the crucial determinant in the differentiation of the three Malayan traditions from this presumed common ancestral community. Each way of making a living imposes different constraints on social organization. Kinship systems, for example, encode 'the preferred ideal pattern of consociation and spatio-temporal distribution' of groups (Benjamin 1986). Optimum group size and structure will differ between foragers and settled farmers (see Chapter 2) leading to differences in kinship rules and other socio-cultural patterns and expectations. Over time, disjunct cultural traditions stressing different kinship composition and interrelationships of local groups developed. Each social pattern structured local population groups in different ways with different consequences for rates of gene flow and magnitudes of genetic differentiation.

As already noted, the subsistence distinctions among the three traditions are Semang-foragers, Senoi-swiddeners, and Melayu Asli-farmer–traders. Table 5.1 lists the associations between group, linguistic affiliation, subsistence mode, and basic societal pattern following Benjamin (1986). Figure 5.4 shows the geographic pattern of these traditions with Semang foragers mainly to the northeast, Senoi along both flanks of the central mountain range of the peninsula, and Melayu Asli more southerly. It

Table 5.1. *The three 'traditions' of Malayan Orang Asli*

Tradition	Language	Technology/ economy	Societal pattern
Semang ('Negritos')	Northern Aslian	Nomadic foragers	Exogamy, mobile conjugal families, extensive networks
Senoi	Central Aslian	Sedentary swiddeners	Nodal kindreds, fision–fusion
Melayu Asli ('Aboriginal Malays')	Southern Aslian	Sedentary farmers, collectors-for-trade	Endogamy

should be noted that the subsistence and societal patterns listed in Table 5.1 do not apply uniformly to present-day Orang Asli populations since many Semang groups have been resettled in villages by the government and most Orang Asli are now engaged in some aspect of the cash economy – small-scale rubber tapping or orchard cropping. Also several large Melayu Asli groups now speak dialects of the Malay language. A further caveat applies to the fixity of these categories. Language, culture, and biology are not absolutely linked entities and some Orang Asli groups combine attributes of different categories. The Lanoh, for instance, are considered to be a Negrito group (Semang), yet they speak a Central Aslian (Senoi) language and share some aspects of Senoi social organization. Over the long term, however, the three traditions appear to have occupied distinct cultural 'niches', specializing in different ways of life, and remaining relatively separated.

Turning now to a closer look at the implications of these different subsistence modes on group composition, a primary division is between nomadic hunter–gatherers (Semang) and settled farmers (Senoi and Melayu Asli). The latter two groups are distinguished primarily by the importance of external trade for the Melayu Asli (Dunn 1975).

The nomadism of the Semang way of life involves constant dispersal and unstable local groups (see Chapter 2 for a comparable pattern in a different environment, the !Kung San). There is little pressure for extended association among kin, the conjugal family is the basic social unit with no larger corporate residential or kin group. Exogamy is extreme including all consanguineal relatives but also extending to affinal kin so that marriages must be made with families without previous ties. The orientation of Semang life, then, is outward. A kin network spread over a wide geographic range allows unrestricted movement in keeping with their foraging subsistence mode.

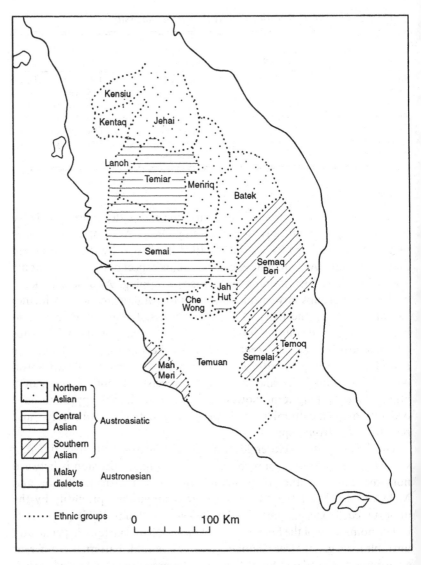

Figure 5.4. Distribution of Orang Asli groups in Peninsular Malaysia. (From Fix 1995.)

Both the Senoi and the Melayu Asli patterns, based on sedentary farming, contrast with the Semang foragers. Sedentism implies a more concentrated residential group and often corporate groups form in order to manage land and task groups. The Senoi system of extensive swidden farming (rather than more intensive wet-rice padi agriculture practiced by most Malays and some Melayu Asli) does not require either the tight control over scarce valuable land itself or the organization of an irrigation system. Senoi settlements are much more stable than Semang camps, however, they are more labile than those of the Melayu Asli. The Semai, described in some detail in Chapter 2, are the largest of the Senoi subgroupings and provide a good example of this pattern. Semai settlements comprise a core group of relatives and a peripheral group of less closely related persons, similar to the 'nodal kindred' discussed by Helm (1968) for the Dogrib Indians of Canada. This is a pattern that allows incorporation into the group through any kin link. For the Semai, most newly married persons move back and forth between the settlements of the husband and wife for several years, finally choosing one for long-term residence. Residence is flexible and changes are relatively easy. Ties between individuals and families in the different settlements are promoted by a prohibition on marriage with consanguineal kin. In contrast to the Semang, Semai prefer affinal kin as marital partners, a practice that increases the density of the kin network locally. Lacking strong corporate groups and leaders, Senoi settlements are prone to fission (see Chapter 2). Fusion with nearby settlements is facilitated by the marital network extending throughout the region.

The Melayu Asli combine sedentary farming with trade. A long history of collecting of forest products for trade with outsiders exists in the peninsula (Dunn 1975). Benjamin (1986) sees this combination as the genesis of the Melayu Asli pattern of localized kin groups based on endogamous marriage. Corporateness of residential groups both facilitates cooperation in agriculture and competitive position with other comparable trading villages. Thus Melayu Asli villages would tend toward greater stability and self-containment (and more closely approximate a local breeding population as well).

These contrasting social arrangements have clear consequences for patterns of intermarriage, mobility, and other aspects of population structure. The Semai Senoi pattern has already been used to illustrate fission and fusion population structure and kin-structured migration (Chapters 2 and 3). Although much more could be said about how these social patterns affect microevolution (Fix 1995), in this chapter I will focus instead on the long-term implications of such social differentiation.

Biological origins of the Orang Asli

If Benjamin's reconstruction of the *in situ* origin of both the linguistic and cultural differences among the three Malayan Orang Asli groups is correct, then the traditional view of biological differences in the Peninsula being the outcome of waves of separate migrations must be false (or at least not completely true). A key component underpinning the migration theory is that the differences are too profound to have arisen locally from a single ancestral population. The relatively short span of 5000 to 6000 years might be insufficient time to evolve a distinctive suite of 'racial' characteristics. However, examination of the actual variation (such as has been documented) shows that biological differences are not neatly apportioned among the three groups. Indeed, one of the major distinguishing features, the 'pygmy' stature of the Semang Negritos, fails to differ significantly from the Senoi. Other characteristics such as hair form and skin color are sufficiently variable within groups to obviate their use as population 'markers'.

The traditional diagnosis of Malayan biological groups was based on the visual traits then considered to mark population differences: skin color; hair form; and various morphological characters such as nose shape and ear form. Each type, then, was defined on the basis of the concurrence of a set of markers. Negritos had dark skin color, 'wooly' hair, and were very short; Senoi (then called 'Sakai', a derogatory Malay term) were somewhat lighter in skin color, had wavy hair, and were taller; Melayu Asli were even lighter in skin color, possessed 'lank' straight hair, and so forth.

The empirical data validating these categories were scanty. Variability within groups was recognized but usually ascribed to admixture. Thus Skeat and Blagdon (1906) described a group of 'Jakun' (Melayu Asli) as obviously mixed with Negrito and Senoi 'blood' since hair form among the group varied from tightly curled to straight. Individuals irrespective of their geographic or populational affiliation were typed by their possession of diagnostic traits. Thus Evans (1915) observed two 'Negrito' (appearing) persons in a Senoi village and reported them to be Semang despite the fact that their reputed homeland was simply another Senoi district. Noone (1936) found some Senoi populations with relatively dark skin color and tightly curled hair, and described them as an 'older stratum' within the Senoi group.

Frequency data are notably lacking. Where these data are reported, traditional typology is not supported. Williams-Hunt (1951), for instance, noted that in nearly 100 Negrito hair samples, he found 'only three true curly bits' with the rest all being medium to deep waves. Similarly, quantitative data on stature, a trait especially thought to distinguish the 'pygmoid' Negritos, show no differences in height between Semang and

Table 5.2. *Stature of Semang and Senoi*

Group	N	Height (cm)	N	Height (cm)
		Males		Females
Senoi (Semai)[1]	48	151.5 (140.5–161.7)	32	142.1 (135.6–156.2)
Senoi[2]		152.0		142.0
Senoi (Semai)[3]	35	158.0		
Senoi (Semai)[4]	50	154.7		
Semang[5]	127	153.6	86	142.7

[1]Fix, Unpublished data; [2]Martin (1905); [3]Polunin (1953) Highland groups; [4]Polunin (1953) Lowland groups; [5]Schebesta (1952).

Senoi. Table 5.2 summarizes these data. The Semai living at higher elevations in the mountains are a bit taller than those living in the lowlands but otherwise there is remarkable consistency in Orang Asli stature. If anything, the large sample of Semang measured by Schebesta (1952) are slightly taller than my sample of lower elevation Semai. Clearly there is nothing in these data to distinguish the groups.

Gene frequency data (summarized in Lie-Injo 1976) are too few to compare among the groups. Again, within group variation is high, particularly for the Senoi for reasons discussed in Chapters 2 and 3. For example, the ABO*O blood group allele frequency varies from 0.432 to 0.840 in sample sizes ranging from 34 to 200 among seven Semai settlements (Fix & Lie-Injo 1975). Molecular data seem no better at resolving distinctions at this level of partition. Ballinger *et al.* (1992) and Melton *et al.* (1995) have analyzed mitochondrial DNA from Orang Asli. The sample in the first study included 33 individuals, most of whom were Senoi although more than half (17) were unidentified as to 'tribal' origin. The Melton *et al.* (1995) samples were all Semai Senoi. The Ballinger *et al.* (1992) protocol used restriction fragment length polymorphisms of mtDNA to characterize a series of Southeast Asian populations including Vietnamese, Malays, Koreans, and Sabah Aborigines. Essentially all that can be concluded is that all these groups share common ancestry. Melton *et al.* (1995) show that the frequency of a specific 9bp deletion (between the COII and tRNA[Lys] mtDNA genes) is quite similar among Semai, Filipinos, and Taiwan aborigines, not a very helpful result for the present concern.

Similar broad scale surveys of other biological characteristics reinforce the notion of a generalized Southeast Asian population. Hanihara (1993) based on cranio-facial morphology, and Turner (1990) from dental morphology, reach the same conclusion as the molecular genetic studies – modern Southeast Asians derive from a common 'late Pleistocene stock'

(Turner 1990:315). History certainly records many large-scale population movements in Southeast Asia but these are not clearly marked by the currently available biological data. This problem of reconstructing historical population affinities in Southeast Asia from genetic data will be considered in more detail below.

Another angle on the question of ancient migrations could be provided by archaeological skeletal samples of the previous occupants of the Peninsula. The migratory wave theory predicts a series of population strata beginning with Negritos; presence of Negrito skeletal traits in early sites would be evidence for the theory. However, Bulbeck (1981) found no evidence for this in a study of pre-Neolithic (ostensibly pre-Senoi) skeletons from Gua Cha, Malaysia (dated between 3000 and 10,000 years ago) representing the Hoabinhian tradition that some consider to be the precursor of the modern Orang Asli (Solheim 1980). Indeed, stature in these populations is greater than for either Semang or Senoi. Other sites in Southeast Asia also fail to show evidence of a Negrito substratum (Bellwood 1993; Bulbeck 1981).

To summarize, none of the existing data strongly supports a wave theory of Malayan Orang Asli origins. On the other hand, these data are not sufficient to disconfirm the hypothesis either. The molecular and morphological (and probably archaeological) data suggesting a broadly distributed ancestral Southeast Asian population would not be inconsistent with migrations from within the region into the peninsula (Bellwood 1993). The point to make is that biological data do not speak for themselves. We have no unambiguous markers of past population affinities labeling individuals or populations. Biological similarity can arise through multiple mechanisms, only one of which is historical common descent. Shared biological characteristics, then, may suggest hypotheses about history. These conjectures must be rigorously tested using data from as many independent sources as possible and careful attention must be given to alternative hypotheses.

Population units in historical analysis

This consideration of the genetic prehistory of the Malayan peninsula simply took as given the traditional division of the Orang Asli into three groups and inquired about their history. Similarly, human geneticists have calculated times of divergence for traditional human racial divisions (e.g., Nei & Roychoudhury 1974) and Cavalli-Sforza *et al.* (1994) present dendrograms purporting to reflect the phylogenetic history of various human groupings. However, as suggested at the beginning of this chapter, it is not certain that traditional 'racial' groups or variously defined ethnolinguistic groups are actually the appropriate units for the analysis of long-term human genetic history. If the arguments occasioned by the Human Genetic Diversity Project (Baer 1993; Roberts 1992) are any indication, fundamental questions about the population units of analysis remain to be answered.

For many small-scale agrarian societies, the deme, defined on the basis of current endogamy, is often a subgroup of an ethnolinguistic unit, the 'tribe'. Thus, as we saw in Chapter 2, clusters of Yanomamo or Semai villages formed relatively endogamous units linked together by common histories and continuing gene flow and differing genetically from other such clusters. Genetic microdifferentiation could be related to the history of village fissions and fusions. Over the longer term, however, gene flow among such units would be sufficient to tie them together into broad regional populations.

Population density and level of socio-cultural integration affect the size of these regions. Low density, mobile, highly exogamous foraging populations form networks of kin ties across wide areas. Practices such as the linguistic group exogamy found in the Vaupés region of Columbia (Chapter 2) and the prescription on marrying any kinsperson, consanguine or affine, of the Semang of Malaysia (above) illustrate this pattern.

In so far as either interlinked village clusters or networks of mobile individuals characterized human breeding structure over the long evolutionary span, the isolated lineages implied by the phylogenetic tree diagrams used to represent genetic history misrepresent the course of human evolution. A reticulate, anastomosing network of populations rather than diverging branches would provide a better model. Of special importance, network models also emphasize the channels by which adaptive genes spread across the species.

The problem of choice of appropriate analytical units for different questions, whether of historical inference or evolutionary process, can be illustrated by the Malayan Orang Asli case. The three traditions coexisting in the peninsula range from foragers to settled agriculturalists and patterns of mobility and local group stability vary among groups. The Semang marriage pattern of extreme exogamy should allow the rapid diffusion of genetic variants among the entire population as in each generation individuals find spouses in distant bands. Local bands should exhibit little genetic differentiation and the entire ethnolinguistic group would be the short-term breeding unit, with mating being biased from random by the avoidance of kin. In contrast, the Melayu Asli preference for community and kin endogamy should produce high levels of local village-level genetic differentiation over the short term. The Senoi pattern is intermediate with localized nodal kindreds forming the core of settlement populations that periodically fission, with groups of relatives hiving off to fuse with other groups or start a new settlement (see Chapter 2 for more details on the Semai Senoi).

The genetic data available for these populations is consistent with these expectations. Although so few data are available for the Semang that little can be said, a study of ABO blood groups by Polunin and Sneath (1953) showed that four Jehai (Semang) settlements were essentially homogeneous. The same study found sharp contrasts among Melayu Asli settlements. In fact, Polunin and Sneath note that these villages 'are virtually small tribes in their own right' (Polunin & Sneath 1953:244). Baer *et al.* (1976) also reported microdifferentiation among four Temuan (Melayu Asli) settlement areas. For the Semai Senoi, Lie-Injo and I (Fix & Lie-Injo 1975) demonstrated sharp differences among settlements for the three loci considered, despite the fact that two of the alleles, hemoglobin E (HbE) and ovalocytosis, were almost surely subject to malaria selection. At the same time, the continual flux introduced by the fission–fusion process would ensure that alleles would be more widely spread over a longer time span (Fix 1975).

In the microevolutionary temporal scale, even within the microcosm represented by the peninsula, the genetic landscape can be seen to be quite complex. What implications does this complexity have for the longer evolutionary time scale? Do these ethnolinguistic groups maintain any genetic cohesiveness for hundreds of generations? To answer this question we might consider various partitions of the human population from Semai villages to the larger Southeast Asian and Oceanic region. Clearly villages will refer only to short-term microevolutionary processes and reflect recent history. As the evolutionary time scale of interest increases, so must the unit of analysis be broadened to include the 'tribe', the language group, all Orang Asli, Southeast Asians, and finally, the species. Furthermore, depending on the demographic and genetic evolutionary processes affecting alleles, different partitions may be more or less useful. Thus if all variation arises by binary fission of populations that then diverge only via genetic drift in isolation, as might be (partially) true for species level evolution, the obvious units to be compared are the isolated fission products. If, on the other hand, there is a continuous nexus of gene flow among all the component local populations of a region or of a species (as postulated, for instance, by the multiregional model of human origins discussed above), and further there are some genes with selective advantage over wide geographic ranges, the trees of 'tribal' or 'racial' populations make less sense.

Looking at everyone's favorite molecule, mtDNA, the picture provided by the previously cited work of Ballinger *et al.* (1992) and Melton *et al.* (1995) shows genetic connections among all the Southeast Asian populations sampled. Figure 5.5 from Ballinger *et al.* (1992) is a genealogy of mtDNA

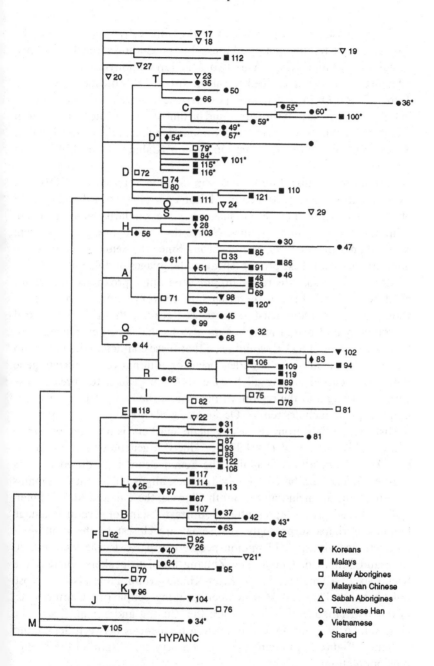

Figure 5.5. Genealogy of Southeast Asian mtDNA haplotypes. Symbols designate ethnic origin of the haplotype (slightly modified according to corrections supplied by the authors). (From Ballinger *et al.*, 1992 fig. 2.)

haplotypes based on 191 polymorphic sites using restriction enzymes . The Orang Asli are represented by 33 individuals including seven Temiar, five Semai, one Jakun (Melayu Asli), two 'Jeni' (an ethnonym with uncertain referent), and 17 of unidentified 'tribal' origin (probably most were Semai or Temiar Senoi). Orang Asli haplotypes have been marked with open squares in Figure 5.5; notice that they are found in almost *every* cluster. This pattern might reflect gene flow across the continental region and/or variation that 'predates the present geographic subdivision' (Ballinger *et al.* 1992) of the region.

Turning to the distribution of a 9bp deletion between the COII and tRNA^{Lys} mtDNA genes studied by Melton *et al.* (1995) based on a sample of 30 Semai Senoi, Figure 5.6 shows a close similarity among Semai, Filipinos, and Taiwan aborigines. Notice that all the branches are deep; rather than a tree, this is a stump sprout. Since the Semai share no close linguistic ties with these groups (speaking languages of different *families* – Austroasiatic versus Austronesian) nor geographic proximity, this result must be interpreted (as in the discussion above) as simply evidence for widespread shared Southeast Asian genes. Recalling that the cranio-facial morphology and dental morphology also fit the same generalized pan-Southeast Asian 'stock', it would seem that these characteristics record no history subsequent to the late Pleistocene or, alternatively, sufficient gene flow has occurred among all Southeast Asian groups to spread these variants throughout the region. In the case of the 9bp deletion, this geographic spread includes remote Oceania as well (Melton *et al.* 1995).

Historical insight from classical genetic 'markers' is no more penetrating. Cavalli-Sforza *et al.* (1994:234) produced a genetic tree of 25 Southeast Asian populations including the Semai Senoi reproduced here as Figure 5.7. The tree, based on only 31 genes, aligns the Semai with mainland populations including the South Chinese, the Tai, and Mon-Khmer, again a broad group representing Southeast Asians of several language families. Perhaps surprisingly, the Semai are closest to the Zhuang, a large ethnic 'minority' (13 million people compared to the less than 20 thousand Semai) of China. The Zhuang are Tai-speakers whereas the Semai speak a language in the Mon-Khmer group. Note also the close links between Thai and Mon-Khmer-speaking people (the Cambodians) of different language families, as are the Zhuang and Semai. Recall the close similarity in mtDNA 9bp deletion frequencies between Semai and Filipinos; Figure 5.7, in contrast, shows a very early 'split' of the Philippine population.

Figure 5.8 presents another tree based on nearly the same number of genes (30) but clustering a different set of populations (Saha *et al.* 1995).

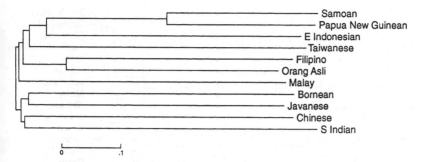

Figure 5.6. Tree based on mtDNA haplotypes and the 9bp deletion for 11 Asian populations. (From Melton *et al.* 1995, fig. 2.) Scale: genetic distance.

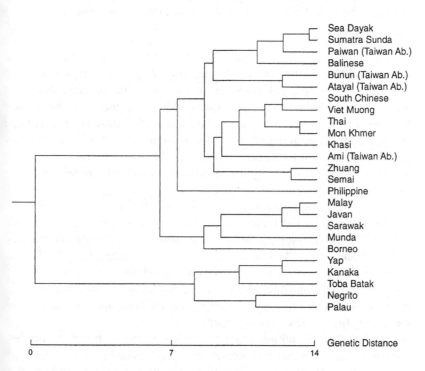

Figure 5.7. Gene tree for 25 Southeast Asian populations based on 31 genes. (From Cavalli-Sforza *et al.* 1994, fig. 4.13.1.)

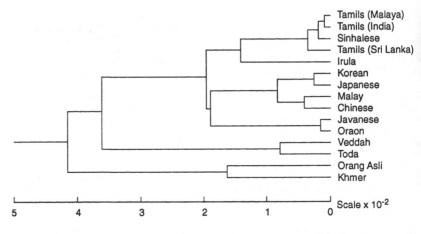

Figure 5.8. Gene tree for 15 Asian populations based on 30 alleles from 7 polymorphic loci. (From Saha *et al.* 1995, fig. 2.) Scale: genetic distance.

Although the population label is 'Orang Asli', the genetic sample is in fact 349 Semai Senoi from Pahang State. In contrast to the previous dendro-gram, this tree clusters the Semai closely with the Khmer, their linguistic relatives. This pair in turn is rather isolated from other clusters. Since Hb E, an allele advantageous in malarious environments, is high in both of these populations and not in the others, it may well be that natural selection is determining the 'history' implied in this fission-and-divergence tree dia-gram.

To interpret these diagrams as anything other than phenetic measures of genetic dissimilarity seems impossible. It may be, as Cavalli-Sforza and his colleagues (1994:235) suggest, that there are simply too few data to generate 'a reasonable genetic taxonomy'. However, when Saha *et al.* (1995) in-creased the number of alleles in their analysis to 53 (reducing the number of populations that could be clustered), the Semai clustered with the Javanese, speakers of an Austronesian language, not an intuitively obvious result and weakening the 'too few data' argument.

The alternative hypothesis to a binary fission populational history in Southeast Asia is the network model of weakly isolated local populations exchanging genes with their neighbors (as the ethnographic record invari-ably shows – see Chapter 2) generation after generation. Over longer periods, populations wax and wane, some expanding while others contract. Rarely, opportunities such as the empty islands of remote Oceania allow colonization and range expansion.

Hemoglobin E, ovalocytosis, and history in Southeast Asia and Near Oceania

The distribution and history of these two genes encapsulate several of the points raised in preceding sections. Both genes protect against malaria and therefore their frequencies in populations are determined by natural selection; in malarious environments we should expect to see high frequencies of both alleles, all other things being equal. Yet as one looks at these gene frequencies among an array of variously sized Southeast Asian populations (Table 5.3), it seems clear that the two genes are to some degree disjunct in distribution. Hemoglobin E is a mainland gene. In fact, Flatz (1967) argued strongly that its initial center of distribution was in the Mon-Khmer speaking peoples from Viet Nam to Burma. As Table 5.3 shows, Hb E frequencies in populations of Thailand are quite high (see also Livingstone 1985). These areas also have been subject to holoendemic malaria and the situation appears to be one where an adaptive variant arises in a population and then diffuses throughout the range in which it has selective value. Notice that the Senoi (including the northern Temiar, and various groups of the central Semai), Mon-Khmer speakers, share this genetic connection with their mainland linguistic relatives. Indeed, something of a cline exists from north to south, with the highest frequency in the northern Temiar, grading through the more northerly Perak Semai, to the south-central Semai of Pahang State. However, a gradient determined by three points is not very convincing and the temptation to produce diffusion or selective gradient explanations should probably be resisted in this case, especially when the small sample size of the Temiar is considered. The very high frequency in the settlement of RU in Pahang (demonstrably caused by a kin-structured fusion event) reinforces this explanatory caution. The Semelai, Southern Aslian-speaking Melayu Asli, continue the north–south trend, however, with a somewhat lower frequency of Hb E. The other two Melayu Asli populations that have been surveyed (Temuan and Jakun) are barely polymorphic at the hemoglobin locus. The Hb E gene frequency for Semang (Lie-Injo 1976) is very low ($q_E = 0.024$) but the sample size was unreported and presumably very small. This relative lack of Hb E (if true) could be due to the low transmission of malaria among these nomadic people (see Fix 1995 for further discussion of this point).

Among lowland populations such as the Malays of Selangor State and throughout many rice-farming areas of Indonesia, malaria endemicity is erratic (Livingstone 1967) and Hb E frequencies are generally low in these populations. An exception is the more malarious inland region of Trengganu State where the frequency in adults is over 10 percent. Whether this

Table 5.3. *Hemoglobin E (Hb E) and ovalocytosis frequencies*[1]

Population	Size	Ovalocytosis		Hb E	
		N	% Positive[2]	N	q_E
Negritos		?	< 2.0	?	0.024
Senoi	39,000				
Temiar	13,000	?	6.7	80	0.319
Semai	19,500				
Perak	12,000	242	6.6	332	0.255
Pahang	7500	545	21.3	520	0.215
SA	272	196	25.2	198	0.168
RU	107	81	28.4	75	0.346
BU	107	79	5.1	80	0.250
Melayu Asli	28,000				
Temuan	10,500	315	35.2	406	0.015
Semelai				41	0.171
Jakun	11,500	?	19.0	116	0.017
Malays	8 × 10⁶				
Selangor		?	< 0.3	536	0.015
Negri Sembilan		629	13.2	629	0.026
Trengganu				249	0.111
Indonesians	190 × 10⁶				
Sulawesi	14 × 10⁶	?	40.0–50.0		
Bali	3 × 10⁶			219	0.018
Minangkabau		83	7.2	235	0.011
Papua New Guinea	4 × 10⁶				
Kar Kar Island		334	13.8		
Thailand					
Khmer				133	0.327

[1] Data from: Amato & Booth 1977; Lie-Injo 1976; Livingstone 1985; see also (Sofro 1986) for more data on Indonesia.
[2] Phenotype frequency.

allele was introduced to the Malays in this region via gene flow from the Orang Asli is unknown. The opportunity surely existed and given the selective advantage of the gene, its relatively high frequency is understandable.

Hemoglobin E frequencies, although influenced by natural selection, are congruent with the linguistic evidence for an historical connection between the Orang Asli and a formerly more widely distributed Austroasiatic-speaking population across mainland Southeast Asia. Ovalocytosis frequencies present the 'other side' of Orang Asli history, the connections to the islands of Indonesia and near Oceania. As will be seen, this genetic history is less by descent and more a result of gene flow and selection. A credible case can be made for the explanation for ovalocytosis in Orang Asli as the complement to the presence of Hb E in Trengganu Malays. It seems likely that this adaptive allele, occurring in high frequency in some island peoples, has been diffusing (the classic wave of advance of an advantageous gene; Fisher 1937) north through the malarious regions of interior Peninsular Malaysia.

The distribution of ovalocytosis extends in a broad swath across Indonesia. Polymorphic frequencies of this dominantly-inherited condition have been found from Papua New Guinea (coastal regions, as well as Kar Kar Island listed in Table 5.3), and in pockets through Indonesia to the Peninsula Malaysia (Livingstone 1985; Sofro 1986). The evidence for the malarial protective function of the phenotype is strong (Baer *et al.* 1976; Serjeantson *et al.* 1977) Since the incidence of malaria is erratic through Indonesia, many populations in the region seem not to possess the allele in polymorphic frequencies; however, where it does occur, it can be in very high frequencies as in Sulewesi where phenotype frequencies of up to 50 percent were recorded (Livingstone 1985). Beyond Peninsula Malaysia, there is no evidence for ovalocytosis in mainland Southeast Asia. Since the ovalocytosis phenotype is an easily observed change in the shape of the red blood cell to an oval or elliptic rather than disk form, it seems unlikely to have been unnoticed if actually present polymorphically.

In the Peninsula of Malaya ovalocytosis frequencies vary from almost absent (among Selangor Malays with no history of malaria) to over 35 percent in the Temuan Melayu Asli (Lie-Injo 1976). Not surprisingly, given the focus of the gene in the islands, the highest Orang Asli frequency is among Melayu Asli. The Negrito Semang frequency, again with an unspecified but surely very small sample size, is very low, consistent with a low malaria incidence. The Senoi present an interesting intermediate population for ovalocytosis. The northern and western Temiar and Perak Semai are very similar in their low polymorphic frequency. Ovalocytosis in Pahang Semai, in contrast, is quite frequent, about three times more common than in Perak.

The situation of a relatively high frequency of ovalocytosis among the Pahang Semai shows us exactly how genetic 'affinity' can result from gene flow rather than descent from a common ancestor. Ethnohistorical information allows us to reconstruct some of the circumstances leading to the introduction of this allele to the Pahang Semai. According to local Semai, several Orang Asli from the south (possibly Temuan from Selangor), escaping from Malay raiders, arrived in the local area and intermarried with Semai. A genealogy of one of these ancestral southerners (Fix 1995) shows that some 18 of the 50 ovalocytosis positive individuals in one settlement are his direct descendants. The relative recency of this flow is evidenced by the fact that the grandchildren of this immigrant were in their late fifties and early sixties in 1968 when this information was recorded. Further evidence on the timing of the event is provided by the ethnicity of the raiders. Semai claimed they were Rawa, a Malay group known to have caused numerous 'disturbances' on the peninsula from the 1850s to about 1870 (Milner 1978). Evans (1915) also notes the presence in another Semai village of descendants of Selangor Orang Asli who had been sold into slavery by Rawa and other Malay raiders and after gaining their freedom, married Semai and produced offspring (see also Dentan *et al.*, 1997, for more details on the effects of Malay incursions on the Orang Asli).

Melayu Asli and Senoi became separate cultural traditions beginning some thousands of years ago according to Benjamin's model. Cultural separation has no doubt limited gene flow to some extent but has not raised insurmountable barriers. Senoi and Melayu Asli are not separate branches on a genetic tree; their evolutionary history is not one of binary fission and isolation. The Semai are connected to the historical gene pool of mainland Southeast Asia and share genes such as Hb E with their Khmer linguistic relatives but they are also connected to Sulawesi and coastal New Guinea through intermarriage with Melayu Asli who maintained trade (and marital) links with the larger Malay trading sphere extending throughout the islands.

6 Conclusions: an evolutionary framework for the study of migration

Migration, and the consequent exchanges of information and genes between populations, is one of the principal forces affecting human diversity, both genetic and cultural. The major focus of this book has been the comparative study of migration (and its converse, isolation) in contemporary human populations and its role in determining levels of genetic microdifferentiation among subpopulations. The framework for this study has been provided by anthropological genetics through the concept of population structure, an approach that integrates biological, demographic, and socio-cultural factors in an ecological framework (Harrison & Boyce 1972a). An important goal has been to identify basic variables affecting the degree and pattern of human mobility and consequently genetic variation. This anthropological approach is microevolutionary, concerned with processes acting over relatively short temporal and spatial scales.

A common assumption from Darwin to the present has been that these microevolutionary processes operating over the long-term are responsible for macroevolution. Thus, the forces of evolution generating gene frequency change in local populations are the same mechanisms responsible for the larger differences characterizing regional or continental populations, small variations having accumulated to larger ones. Specifically, patterns of migration observed in modern populations, augmented by dispersion and colonization of new ranges through history, have produced the pattern of contemporary global human diversity. On this view, then, understanding how migration works in present day populations should also inform us about the longer term evolution of our species.

However, as Harpending et al. (1996) point out, this anthropological tradition of studying local process and variation (that they label the 'population perspective') is not shared by many geneticists concerned with a longer time span of evolution and the accumulation of neutral mutations leading to species differentiation (a perspective they call 'phylogenetic').

While in principle, phylogeny simply refers to the evolutionary history of groups or lineages, in practice, phylogenetic classifications represent the genesis of diversity exclusively as a branching process as populations split,

203

disperse, and become isolated. Genetic divergence occurs in isolation, therefore, the degree of genetic difference between groups is a measure of their recency of common ancestry. Evolutionary history is a sequence of binary fissions. From this point of view, local genetic differentiation is simply 'ripples on the gene pool' (Harpending & Ward 1982). Global patterns of diversity owe little to population structure and the study of contemporary migration patterns and population models of genetic variation are replaced by branching tree diagrams of diverging gene pools (e.g., Cavalli-Sforza *et al.* 1994).

The phylogenetic approach is nearly ubiquitous in current analyses of molecular variation. Dendrograms purporting to show the history of population splitting are the preferred form of presentation of such data. The problem with this approach is that human populations are not reproductively isolated species and therefore tree diagrams may often be inappropriate representations of their evolutionary histories. Populations may fission, be geographically separated but continue to exchange genes, and even fuse together again at a later date. Because of this fact, reticulate, anastomosing models of evolution may be better descriptors of the history of populations than are trees (Moore 1994).

On the other hand, the equilibrium models of classical population genetics, assuming constancy of migration pattern and rate over long time periods, as we saw in Chapter 3, strain credulity. Undoubtedly, major population rearrangements have occurred over the long term, sometimes leading to isolation and divergence of the daughter populations.

Neither perspective is likely to supply all the answers. Surely Harpending *et al.* (1996:283) are correct in stating that '[O]ur current interest in global human origins and dispersions calls for a blend of these traditions both in theory and in techniques of data analysis and presentation', furthermore, '[T]he study of modern human origins and the study of human variation are the same, since they both refer to processes by which contemporary human diversity developed'.

One way to facilitate this blending would be to recognize that the population and phylogenetic perspectives are the ends of a temporal and spatial continuum. While the extreme branching model applies only to the history of species or genes, in some ecological and historical contexts, fission and isolation may have been more typical of populations than the localized exchange of mates commonly found in contemporary small-scale human societies. Current and recent historical human societies have differed widely in mobility patterns and some, such as the Yanomamo of Amazonia, give some insight into how splitting and divergence might occur. A general evolutionary theory of migration, encompassing the entire range of potential movement from range expansion, fission, and dispersion to extreme philopatry, would specify the appropriate population units and mobility patterns for the social and ecological context and temporal and spatial scales of analysis.

The connection between the study of contemporary migration and genetic variation and the long-term evolution of our species must be in the development of such a general theory of the causes and patterns of movement. This can not be simply ethnographic analogy ('!Kung in the Pleistocene' models) but rather a search for basic variables affecting the process of migration – that is, *conceptual models* in the sense of Tooby and DeVore (1987). Clearly no present-day population is identical to any past population, just as all species are unique (Foley 1987). Just as clearly, the uniqueness of each species does not preclude a comparative biology. Similarly, a comparative perspective on the ethnography of human movement is a crucial source of hypotheses about past patterns.

There are no living fossils but by the same token no fossils are alive. Direct evidence of behavioral patterns, including migration, of past populations is limited to inferences from the material remains in the fossil and archaeological records. Although archaeologists and paleontologists have devised many ingenious techniques to extract information from this record, such inferences can never be as detailed as the description possible with living populations. This inherent limitation on our knowledge of ancient populations increases the importance of evolutionary ecological models of the migration process.

To apply these models to past populations depends on identifying archaeological signatures for the relevant variables such as population density. Indeed, careful ethnoarchaeological studies of contemporary populations by archaeologists, such as that of Yellen (1977), have provided some of these indicators.

Behavioral ecology of migration

The development of a general evolutionary theory of migration might begin with behavioral ecology. Migration is one strategy among several that organisms employ to improve their environmental conditions, both in the somatic (or economic) sense, increasing food supplies and/or decreasing risks of competition or predation (or war), and the reproductive sense, obtaining mates. The frequency and distance of migration will thus depend on the costs and benefits of migration constrained by environmental conditions and the availability of alternative strategies.

In a broad sense, human populations are no different than other species in being constrained by these ecological trade-offs. However, their more complex socio-cultural and economic life adds new dimensions to human behavior.

Causal theory in human migration studies, particularly in the social sciences such as demography and geography, historically has concentrated on economic and spatial determinants. These models emphasize the economic motivations for migration with travel costs associated with distance being the primary constraint on movement, although social variables such as information about potential destinations have also been included. Anthropology, and by extension, anthropological genetics, has long been interested in post-marital residence rules, making marital migration a focus of study.

Anthropology of human migration

The anthropology of human migration, however, should encompass the broader ecological dimensions of migration and mobility. This book surveys this diversity through a series of ethnographic case studies including representatives from the three levels of socio-cultural integration in Johnson and Earle's (1987) classification of social systems: family-level groups; local groups; and regional polities. Three variables define these levels. The basic variable is intensity of economic land use, varying from low intensity foraging to high-investment, irrigated agriculture. The second variable, population density, is strongly correlated with intensity of land use, and the third, social stratification, is a consequence of increases in both the preceding variables.

While not predicting mobility perfectly, clear patterns emerge from these comparisons suggesting that the basic variables of the Johnson and Earle classification are also important determinants of migration. The intuitively obvious relationship between population density and mean marital distance is apparent along the continuum of societies from foragers to intensive farmers. Low density foragers and swiddeners find mates from a much larger area than do high population density, intensive land use societies. Particularly as population density increases, local potential mate pools become larger and the costs of travel to find a marriage partner are avoided. Further contributing to long mating distances of foragers, who lack political integration beyond the family, is the societal strategy of minimizing economic risk by extending affinal kin networks. Marrying 'far' provides kin in other areas who can be 'visited' in times of need. Mobility is also increased in these societies by the lack of political means of resolving disputes – the ultimate response to irreconcilable conflict is fission and flight of one of the parties. High density populations, making intensive use and high investments in land, have a greater commitment to locality. Marriage may serve as a strategy to control land for kin groups and is focused on local ties to retain land. Increasing levels of social stratification may produce alternative strategies for different components of the society. Peasant economies, subsumed within the boundaries of states, may include two or more classes (or castes in the case of India), one of which inherits the land and means to carry on a very intensive, localized agrarian system whereas their non-inheriting sibs (or otherwise non-farming classes) may be extremely mobile in search of economic opportunities outside agriculture.

At least some of this wide diversity of human social arrangements among modern human populations must extend far back in human history and prehistory as differences in land use patterns and population densities reflecting both differing ecologies and technologies occurred. Because of this diversity, there has never been a single 'real' population structure characterizing all humans throughout the long span of our evolution. The considerable diversity among extant hunter–gatherers, ranging from high-ly mobile, thinly spread desert foragers to the dense, sedentary populations of coastal fisher–gatherers, was only augmented by the development of varying agricultural systems. The task, then, is not to rely on 'hunter–gatherers' or any single representative human population as analogies applicable to all human populations throughout prehistory but rather to formulate a conceptual model making use of the relationships between ecological, economic, and social variables and mobility. In so far as these variables are identifiable in the archaeological record, we may gain a more realistic understanding of the role of migration in the development of human diversity.

Population genetic models of migration

The detailed anthropological study of contemporary migration patterns has also played an important role in assessing the assumptions of the classic migration models of population genetics. In order to achieve gener-ality of applicability to all species and mathematical simplicity, these models minimize the number of variables considered. The island model of Sewall Wright, for instance, assumes a very simplified pattern of exchange among populations and his isolation by distance model essentially reduces the determinants of migration to geographic distance.

The relative ease of collecting demographic information on human populations has allowed the development of more complex migration models. The migration matrix model uses the actual migration data on specific populations to predict genetic variation for a region. This information is often available for human populations in the form of parent–offspring birthplace data, whereas comparable data for many other species are more difficult to obtain. Even more explicitly behavioral, and again depending on detailed demographic information, is the neighborhood knowledge model that relates marital migration to visiting frequencies of local populations.

Beyond proposing these more complex models of population structure, anthropological study yields social and kinship data and historical records (sometimes spanning several generations) that allow the identification and evaluation of basic variables in migration as well as the ranges of variation in these variables among societies. Thus one important aspect of migration is the stage of the life cycle at which it occurs: pre-marital; marital; or post-marital. Genetic models are concerned with intergenerational gene flow and ignore these differences in the timing of migration. However, if most dispersal occurs after the high mortality stage of the life cycle, which is the case in the widespread human pattern of marital or post-marital movement, the standard predictions of migration models may not hold. When the dispersing group is small, the migrants constitute a statistical sample of the parental gene pool. As Rogers (1988) has pointed out, under these circumstances, migration is formally equivalent to a component of genetic drift. Rather than homogenizing genetic variation among exchanging populations, such stochastic migration can actually increase gene frequency variance.

Similarly, the classic migration models do not specify the units of migration but subsume the scale of migration under the value of m, the migration *rate*, irrespective of whether individuals or groups do the traveling. These models fail to consider the potential effects of the scale of migration for the structure of migration. Human populations that periodically fission or split off groups that migrate to fuse with other populations differ from the implicit model of random individuals migrating between demes. As for the life cycle stage of migration, migration of groups opens the door for stochastic effects on genetic variation.

In particular, migrant groups can be structured along kin lines as when families migrate together. Since biological kin are likely to share many genes in common, this pattern of migration can lead to highly biased genetic samples of the donor population gene pool. Depending on the degree of relatedness of group members, kin-structured migration may augment rather than reduce differences between donor and recipient populations. Indeed, when stochastic effects are shared (or negatively shared) (Epperson 1994), dramatic effects on spatial patterns of genetic variation can result.

Emphasis in the classic models is on equilibrium patterns of genetic variation achieved after many generations of constant rates and patterns of migration. Population sizes of donor and recipient populations are assumed to remain constant and gene flow must be balanced among groups to maintain this constancy. The migration of interest is local; long distance migration serves in these models as a constant, stabilizing force.

These simplifying assumptions are often violated. The history of human populations records growth and decline through time, changing migratory patterns, and large-scale movements of peoples and trade contacts between distant locales. Gene flow resulting from these long distance contacts may have been crucial in the spread of adaptive alleles such as the hemoglobin variants (Livingstone 1989). Sustained growth can lead to range expansion involving the movement of colonists into unoccupied territory. This 'migratory' process can have numerous genetic effects ranging from founder effect (or the augmented founder effect called 'lineal effect' – Neel & Salzano 1967) to the process Cavalli-Sforza *et al.* (1993) have dubbed 'demic diffusion' where invaders absorb the previous inhabitants of a territory.

More complex models – computer simulation

The more complex population structures implied by relaxing the simplifying assumptions of traditional population genetics models can be studied through computer simulation, which provides a methodology for experimenting with more variables and interactions among variables and processes. Algorithms for generating random numbers also allow stochastic migration and genetic drift to be modeled. Since for much of human evolution, populations were small and kinship was undoubtedly important in structuring groups, migration was likely to have been kin-structured. Range expansion almost surely was by small groups of kin extending their foraging to new areas. Thus, the ability to model non-equilibrium, stochastic processes through Monte Carlo computer simulation allows us to test more realistic scenarios. These models may also help to build a bridge between the microevolutionary study of local genetic variation and that of more global processes of diversification.

Thus, the conclusion that migration always reduces variation among exchanging populations follows from the assumption of the classic models that migrant gene frequencies are identical to those of the population of origin. Under such a deterministic regime, low rates of migration are sufficient to homogenize regional gene frequencies. However, simulation experiments incorporating stochastic migration of kin-structured groups show that high rates of movement are not incompatible with the maintenance of considerable genetic microdifferentiation.

Although this variation is sometimes dismissed as 'ripples on the gene pool', large stochastic effects in population structure have important implications for genetic evolutionary theory. Geneticists attempt to infer evolutionary process or history from gene frequency distributions. If the assumptions of the classic models are accepted, expectations for the level of genetic variation will be incorrect. For example, simulation studies showed that discriminating between neutral and selected alleles based on the expected pattern of their spatial distributions is much more difficult under kin-structured migration. Other experiments showed that kin-structuring of migrant groups can have effects on the rate of spread of an advantageous gene.

Simulation is particularly suited to studying complex population structures through time. Rather than the invariant populations and migration rates of the classic models, time-dependent processes of changing population sizes, range expansions, and stochastic migration can be modeled. Thus, one of the most salient features of gene frequency distributions in Europe is a southeast to northwest cline. Various explanations for this gradient have been proposed, one of which involved a series of repeated founder effects as Neolithic farmers moving from the center of agriculture in the Near East displaced the prior foragers inhabiting the continent (Barbujani *et al.* 1995). Comparison of simulated and observed European gene frequencies showed a good match with the founder effect model. One problem, however, was the very small size of founder groups (often only eight persons), too small to constitute a viable social unit. Further experiments adding kin-structuring of founder groups to the model demonstrated that larger numbers of kin (on the order of 25 persons) sufficient to form a colony, could also replicate the empirical spatial pattern of European gene frequencies. In this case, kin-structure augments the stochastic effects of random founder effect.

If such founder colonies are to establish themselves as viable endogamous populations, the number of potential mates available for individuals must be maintained above some minimum value. The requirement that daughter populations become isolated after fission is critical for the branching models of population divergence of phylogenetic analysis. However, a shortage of potential marriage partners within the population would require continued marital exchange with other populations. Several factors can influence the mate supply including demographic structure and various marriage proscriptions (MacCluer & Dyke 1976). Simulation experiments suggest that very small populations (less than 100 persons) are incapable of maintaining endogamy.

Simulation provided a method to investigate an even more complicated population structure, the structured deme model of D.S. Wilson (1980). This model subdivides the 'deme' into a set of 'trait groups' which are units of natural selection. Trait group selection, as for individual selection, depends on genetic variation among units to operate. Trait groups may differentially grow and their members disperse to other demes thus increasing their genetic contribution to subsequent generations or in the model considered here, some trait groups may be more likely to succumb to disease, surviving groups thereby differentially adding their genes to the next generation's gene pool. Kin-structuring of trait groups may increase variability among groups increasing the potential for selection.

This model approximates the hierarchical structure of Semai Senoi populations in which families are aggregated into hamlets, settlements, clusters of settlements, up to and including the entire population. Hamlets, comprising 'dilute' kin groups, were taken to be the trait groups. Settlements made up of four hamlets were the next level in the hierarchy which, in turn, exchanged migrants with other settlement populations. The rationale for group effects on survival was that hamlets experience more or less epidemic disease mortality depending on the presence or absence of resistant members.

The effect of this population structure on the rate of increase and spread of an adaptive allele was striking, suggesting that under these conditions, evolution can occur very rapidly.

Migration and colonization in the long term

Empirical studies have documented the wide range of migration and mobility in human populations. A great diversity of demographic regimes and socio-cultural traditions are represented in these studies allowing crucial variables structuring migration to be identified. These variables range from broad determinants such as population density, intensity of land use, and level of socio-cultural integration to detailed factors such as the timing of movement and the structure of migrant groups. This wealth of information provides the potential for developing a general evolutionary model specifying rates and patterns of migration. Such a model, combined with archaeological signatures for the variables, would allow us to reconstruct the role of migration over the longer span of human evolution.

However, this great diversity of migratory patterns in the ethnographic and historical records, along with the certainty that not all possible patterns are represented, must also serve as a caution in the reconstruction of evolutionary history from current genetic distributions. As the Man the Hunter conference (Lee & DeVore 1968) demonstrated, there is no single hunter–gatherer population structure; Australians or !Kung cannot provide a model for all pre-agricultural populations. Neolithic Europeans were not equivalent to present-day farmers.

Further complicating historical inference is the fact that alternative evolutionary mechanisms may generate identical genetic distributions. Rules of thumb such as 'natural selection is locus-specific' or 'gene flow always homogenizes differences among exchanging populations' may often not hold. In this book, three examples of current controversies in human historical genetics illustrated the difficulty of unambiguously writing the history of populations from gene genealogies or frequencies.

The first example, the origin and spread of modern *Homo sapiens*, is currently a major focus of debate in human evolutionary studies. The two poles of the controversy are the multiregional model (Wolpoff *et al*. 1994), which posits a long-term geographic subdivision of a dispersed human species population locally adapted to differing environments but connected by gene flow, and the 'Out of Africa' scenario (Cann *et al*. 1987) based on the spread of *Homo sapiens* from its origin in Africa to the rest of the globe, replacing the previous hominid inhabitants. More recently, the 'weak Garden of Eden' scenario (Harpending *et al*. 1993) has been proposed, involving partially isolated subpopulations of the original species population expanding long after the initial speciation event to colonize the rest of the world.

Genetic evidence has played and continues to play a major role in the argument and many would now agree that this evidence has been instrumental in establishing a relatively recent origin of our species. However, despite the impressive fit of the expansion–replacement models to the genetic evidence, unanimous acceptance of any one of these hypotheses has not been achieved. Questions remain about the genetic mechanisms and interpretations but an important unresolved issue is the population dynamics of growth, invasion, and replacement by early humans. Again, a general theory of human migration could help resolve some of these questions and constrain interpretations based on gene distributions.

The second example, the demic diffusion model of Cavalli-Sforza and colleagues (1993), also depends on population growth and expansion. The initial application of the model was to explain the spread of agriculture throughout Europe by expanding populations of Near Eastern farmers but it has also been applied to other regions including Southeast Asia (Bellwood 1996). An alternative hypothesis is that the crops and techniques comprising the European agricultural complex diffused from group to group without necessarily involving movement of people.

The primary genetic evidence supporting the demic expansion of farmers is the apparent cline in gene frequencies across Europe in the direction of putative spread of agriculture. The argument is that the farther the distance from the center of origin of agriculture, the more dilute the genetic contribution of farmers, accounting for the gradient in gene frequencies. Cultural diffusion, in contrast, would ostensibly leave no residue of genes in the population. However, as is often true of models, several alternatives fit the data equally well. The founder effect model, where farmer populations expand and completely replace foragers, has already been mentioned in the discussion of simulation modeling. It is also possible to make a strong case for natural selection as the cause of the clinally distributed genes. A model postulating changing selective conditions diffusing as a result of the borrowing of agriculture was shown to produce exactly the same clinal gradient as the population movement models.

A potential key to testing these alternatives lies in the population dynamics of European prehistoric foragers and farmers. Once again, a strong conceptual model of migration in relation to archaeologically recognizable variables would provide the tools to discriminate between the hypotheses.

The final illustration examines the arguments for migration as an explanation for biological diversity in Peninsular Malaysia. The indigenous populations of the peninsula are divided into three traditions, the foraging Semang (often called 'Negritos'), the swidden-farming Senoi, and the farmer–trader Melayu Asli. The long-accepted explanation for this diversity was separate waves of migration at different times in prehistory. The foraging Semang were considered an ancient substratum widely distributed across Southeast Asia, now persisting in only a few small refuges; the Senoi were thought to be later immigrants as expanding agriculturalists entering the peninsula from the north; and Melayu Asli ('aboriginal' Malays) were even later migrants coming by sea from the south. Each wave was thought to have displaced some of the prior inhabitants.

As an alternative hypothesis, Geoffrey Benjamin (1986) pointed out that all of these groups speak languages that could be reconstructed as descendants of a relatively recent (within the last few thousand years) ancestor and, therefore, all three traditions could have developed *in situ* rather than being already differentiated immigrants from elsewhere. He was able to show

how the linguistic and socio-cultural evidence is consistent with the local differentiation of each group in opposition to the others.

The supposed extreme biological differences among the three subgroups was long taken to be evidence for their ancient and separate origins. However, actual data on morphology and genetics do not support these sharp biological distinctions. For instance, stature of the 'pygmoid' Negritos is indistinguishable from Senoi populations. Mitochondrial DNA patterns suggest ancient connections of Malayan peoples across Southeast Asia while malarial adaptive alleles such as hemoglobin E and ovalocytosis show connections with both the mainland and the islands of Indonesia and near Oceania. Rather than a history of fission, migration, and isolation for these traditions, a broader nexus of gene flow seems to have united all three with the larger regional gene pool including trade patterns with strong effects on the direction and distance of flow.

Population genetics theory provides a powerful tool for understanding evolution. Armed with this theory, geneticists attempt to infer causes from patterns observed in gene distributions. The fundamental problem with this approach is that many different models may fit the same data (Weiss 1988). Depending on the assumed magnitudes of migration, natural selection, mutation, and genetic drift, very different conclusions may be drawn, as in the case of genetic clines across Europe that may be equally well replicated by models of demic diffusion, founder effect, or natural selection.

The solution to this problem is to *estimate* rather than assume the values of the model variables. For migration in contemporary populations (or historical population with records), such estimates can be obtained directly. However, for populations lacking direct documentation, an explicit theory of ecological and socio-cultural determinants of mobility would allow assumptions in particular models to be justified or rejected. A sample of the wealth of information on human populations relevant to the development of such a theory has been presented in this book and key variables such as land use intensity, population density, and level of socio-cultural integration have been outlined. If a general theory has yet to be achieved, sufficient understanding of the process of migration now exists to inform our genetic models.

References

Abelson, A.E. (1976) Population structure in a Pyrenean village: the effects of social class assortative mating and of migration. PhD: The Pennsylvania State University.

Abelson, A.E. (1978) Population structure in the Western Pyrenees: social class, migration and the frequency of consanguineous marriage, 1850–1910. *Annals of Human Biology* **5**: 165–78.

Aberle, D.F., Bronfenbrenner, U., Hess, E.H., Miller, D.R., Schneider, D.M. & Spuhler, J.N. (1963) The incest taboo and the mating patterns of animals. *American Anthropologist* **65**: 253–65.

Adams, J. & Kasakoff, A.B. (1976) Factors underlying endogamous group size. In: C.A. Smith (ed.) *Regional Analysis*, vol 2, *Social Systems*, pp. 149–73. New York: Academic.

Akazawa, T. & Szathmary, E.J.E. (eds.) (1996) *Prehistoric Mongoloid Dispersals*. Oxford: Oxford University Press.

Amato, D. & Booth, P.B. (1977) Hereditary ovalocytosis in Melanesians. *Papua New Guinea Medical Journal* **20**: 26–32.

Ammerman, A.J. & Cavalli-Sforza, L.L. (1971) Measuring the rate of spread of early farming in Europe. *Man* **6**: 674–88.

Ammerman, A.J. & Cavalli-Sforza, L.L. (1984) *The Neolithic Transition and the Genetics of Populations in Europe.* Princeton: Princeton University Press.

Anthony, D.W. (1990) Migration in archeology: the baby and the bathwater. *American Anthropologist* **92**: 895–914.

Aoki, K. (1982) A condition for group selection to prevail over counteracting individual selection. *Evolution* **36**: 832–42.

Aoki, K. & Shida, M. (1996) A Monte Carlo simulation study of coalescence times in a successive colonization model with migration. In: T. Akazawa & E.J.E. Szathmary (eds.). *Prehistoric Mongoloid Dispersals*, pp. 66–77. Oxford: Oxford University Press.

Arensberg, C. & Kimball, S.T. (1968) *Family and Community in Ireland*. Cambridge, MA: Harvard University Press.

Avise, J.C., Neigel, J.E. & Arnold, J. (1984) Demographic influences on mtDNA lineage survivorship in animal populations. *Journal of Molecular Evolution* **20**: 99–105.

Baer, A., Lie-Injo, L.E., Welch, Q.B. & Lewis, A.N. (1976) Genetic factors and malaria in the Temuan. *American Journal of Human Genetics* **28**: 179–88.

Baer, A.S. (1993) Global survey of human genetic diversity: a focal point for human biology. *Human Biology* **65**: 7–9.

Bahuchet, S. & Guillaume, H. (1982) Aka-farmer relations in the northwest Congo Basin. In: E. Leacock & R. Lee (eds.). *Politics and History in Band Societies*, pp.

215

189–211. Cambridge: Cambridge University Press.

Bailey, R.C., Head, G., Jenike, M., Owen, B., Rechtman, R. & Zechenter, E. (1989) Hunting and gathering in tropical rain forest: is it possible? *American Anthropologist* **91**: 59–82.

Ballinger, S.W., Schurr, T.G., Torroni, A., Gan, Y.Y., Hodge, J.A., Hassan, K., Chen, K.-H. & Wallace, D.C. (1992) Southeast Asian mitochondrial DNA analysis reveals genetic continuity of ancient Mongoloid migrations. *Genetics* **130**: 139–52.

Barbujani, G., Sokal, R.R. & Oden, N.L. (1995) Indo-European origins: a computer-simulation test of five hypotheses. *American Journal of Physical Anthropology* **96**: 109–32.

Barker, G. (1985) *Prehistoric Farming in Europe.* Cambridge: Cambridge University Press.

Bayliss-Smith, T. (1994) Melanesian interaction at the regional scale: spatial relationships in a fluid landscape. In: A.J. Strathern & G. Stürzenhofecker (eds.). *Migration and Transformations: Regional Perspectives on New Guinea*, pp. 295–311. Pittsburgh: University of Pittsburgh Press.

Bellwood, P. (1993) Cultural and biological differentiation in Peninsular Malaysia: the last 10,000 years. *Asian Perspectives* **32**: 37–60.

Bellwood, P. (1996) Early agriculture and the dispersal of the southern Mongoloids. In: T. Akazawa & E.J.E. Szathmary (eds.). *Prehistoric Mongoloid Dispersals*, pp. 287–302. Oxford: Oxford University Press.

Bengtsson, B.O. (1978) Avoiding inbreeding: at what cost? *Journal of Theoretical Biology* **73**: 439–44.

Benjamin, G. (1976) Austroasiatic subgroupings and prehistory in the Malay Peninsula. In: P. Jenner, L.C. Thompson & S. Starosto (eds.). *Austroasiatic Studies*, pp. 37–128. Honolulu: University of Hawaii Press.

Benjamin, G. (1985) In the long term: three themes in Malayan cultural ecology. In: K. Hutterer & T. Rambo (eds.). *Cultural Values and Tropical Ecology in Southeast Asia*, pp. 219–78. Ann Arbor: Center for South and Southeast Asian Studies.

Benjamin, G. (1986) Between isthmus and islands: reflections on Malayan palaeosociology. *Working Paper No. 71.* Singapore: Sociology Department, University of Singapore.

Birdsell, J.B. (1958) On population structure in generalized hunting and collecting populations. *Evolution* **12**: 189–205.

Birdsell, J.B. (1968) Some predictions for the Pleistocene based on equilibrium systems among recent hunter–gatherers. In: R.B. Lee & I. DeVore (eds.). *Man the Hunter*, pp. 229–40. Chicago: Aldine Publishing Co.

Birdsell, J.B. (1973) A basic demographic unit. *Current Anthropology* **14**: 337–56.

Bischof, N. (1975) Comparative ethology of incest avoidance. In: R. Fox (ed.). *Biosocial Anthropology*, pp. 37–67. New York: John Wiley and Sons.

Bittles, A.H. & Makov, E. (1988) Inbreeding in human populations: an assessment of the costs. In: C.G.N. Mascie-Taylor & A.J. Boyce (eds.). *Human Mating Patterns*, pp. 153–67. Cambridge: Cambridge University Press.

Bodmer, W.F. & Cavalli-Sforza, L.L. (1968) A migration matrix model for the study of random genetic drift. *Genetics* **59**: 565–92.

Bodmer, W.F. & Cavalli-Sforza, L.L. (1974) The analysis of genetic variation using migration matrices. In: J.F. Crow & C. Denniston (eds.). *Genetic Distance*, pp. 45–61. New York: Plenum.

Boren, T., Falk, P., Roth, K.A., Larson, G. & Normark, S. (1993) Attachment of *Helicobacter pylori* to human gastric epithelium mediated by blood group antigens. *Science* **262**: 1892–5.

Borella, F.T. (1994) *The Genetic Implications of Landowner Class Endogamy in a Spanish Basque Village*. Ph.D. University of California-Riverside.

Boserup, E. (1965) *The Conditions of Agricultural Growth*. Chicago: Aldine.

Bowcock, A.M., Kidd, J.R., Mountain, J.L. *et al.* (1991) Drift, admixture, and selection in human evolution: A study with DNA polymorphisms. *Proceedings of the National Academy of Sciences* **88**: 839–43.

Boyce, A.J. (ed.) (1984) *Migration and Mobility: Biosocial Aspects of Human Movement*. London: Taylor & Francis.

Boyce, A.J., Kuchemann, C.F. & Harrison, G.A. (1967) Neighbourhood knowledge and the distribution of marriage distances. *Annals of Human Genetics* **30**: 335–8.

Boyd, W.C. (1963) Four achievements of the genetical method in physical anthropology. *American Anthropologist* **65**: 243–52.

Bronson, B. (1977) The earliest farming: demography as cause and consequence. In: C.A. Reed (ed.). *Origins of Agriculture*, pp. 23–48. The Hague: Mouton.

Bulbeck, F.D. (1981) Continuities in Southeast Asian Evolution Since the Late Pleistocene. Master's thesis, Australian National University, Canberra.

Cane, S. (1990) Desert demography: a case study of pre-contact Aboriginal densities in the Western Desert of Australia. *Oceania Monographs* **39**: 149–59.

Cann, R.L., Stoneking, M. & Wilson, A.C. (1987) Mitochondrial DNA and human evolution. *Nature* **325**: 31–6.

Cannings, C. & Cavalli-Sforza, L.L. (1973) Human population structure. *Advances in Human Genetics* **4**: 105–71.

Carey, I. (1976) *Orang Asli: The Aboriginal Tribes of Peninsular Malaysia*. Kuala Lumpur: Oxford University Press.

Carneiro, R. (1970) A theory of the origin of the state. *Science* **169**: 733–8.

Cavalli-Sforza, L.L. (1963) The distribution of migration distances: models and applications to genetics. In: J. Sutton (ed.). *Human Displacements*, pp. 139–58. Paris: Hatchette.

Cavalli-Sforza, L.L. (1969) 'Genetic drift' in an Italian population. *Scientific American* **212**: 30–7.

Cavalli-Sforza, L.L. (1984) Isolation by distance. In: A. Chakravarti (ed.). *Human Population Genetics: The Pittsburgh Symposium*, pp. 229–47. New York: Van Nostrand Reinhold.

Cavalli-Sforza, L.L. (ed.) (1986) *African Pygmies*. Orlando: Academic Press.

Cavalli-Sforza, L.L. & Bodmer, W.F. (1971) *The Genetics of Human Populations*. San Francisco: Freeman.

Cavalli-Sforza, L.L. & Cavalli-Sforza, F. (1995) *The Great Human Diasporas: The History of Diversity and Evolution*. Reading, MA: Addison-Wesley.

Cavalli-Sforza, L.L. & Hewlett, B. (1982) Exploration and mating range in African Pygmies. *Annals of Human Genetics* **46**: 257–70.

Cavalli-Sforza, L.L., Menozzi, P. & Piazza, A. (1993) Demic expansions and human evolution. *Science* **259**: 639–46.

Cavalli-Sforza, L.L., Menozzi, P. & Piazza, A. (1994) *The History and Geography of Human Genes*. Princeton: Princeton University Press.

Cavalli-Sforza, L.L. & Piazza, A. (1975) Analysis of evolution: evolutionary rates, independence and treeness. *Theoretical Population Biology* **8**: 127–65.

Chagnon, N.A. (1968) *Yanomamo: The Fierce People*. New York: Holt, Rinehart and Winston.

Chagnon, N.A. (1972) Tribal social organization and genetic microdifferentiation. In: G.A. Harrison & A.J. Boyce (eds.). *The Structure of Human Populations*, pp. 252–82. Oxford: Clarendon Press.

Chagnon, N.A. (1988) Life histories, blood revenge, and warfare in a tribal population. *Science* **239**: 985–92.

Chagnon, N.A., Neel, J.V., Weitkamp, L., Gershowitz, H. & Ayres, M. (1970) The influence of cultural factors on the demography and pattern of gene flow from the Makiritare to the Yanomama Indians. *American Journal of Physical Anthropology* **32**: 339–50.

Chepko-Sade, B.D. & Halpin, Z.T. (eds.) (1987) *Mammalian Dispersal Patterns*. Chicago: University of Chicago Press.

Childe, V.G. (1958) *The Prehistory of European Society*. Harmondsworth: Penguin.

Clark, W.A.V. (1986) *Human Migration*, Scientific Geography Series **7**. Beverly Hills: Sage.

Cliff, A.D. & Ord, J.K. (1981) *Spatial Processes*. London: Pion.

Cole, F.C. (1945) *The Peoples of Malaysia*. New York: Van Nostrand.

Coleman, D.A. (1977) The geography of marriage in Britain (1920–1966). *Annals of Human Biology* **4**: 101–32.

Coleman, D.A. (1979) A study of the spatial aspects of partner choice from a human biological viewpoint. *Man* **14**: 414–35.

Crawford, M.H. & Mielke, J.H. (eds.) (1982) *Current Developments in Anthropological Genetics*. Vol. 2. *Ecology and Social Structure*. New York: Plenum.

Crosby, A.W. (1986) *Ecological Imperialism: The Biological Expansion of Europe, 900–1900*. Cambridge: Cambridge University Press.

Crow, J.F. & Kimura, M. (1970) *An Introduction to Population Genetics Theory*. New York: Harper and Row.

De Jong, G.F. & Gardner, R.W. (eds.) (1981) *Migration Decision Making: Multidisciplinary Approaches to Microlevel Studies in Developed and Developing Countries*. New York: Pergamon Press.

Dennell, R.W. (1992) The origins of crop agriculture in Europe. In: C.W. Cowan & P.J. Watson (eds.). *The Origins of Agriculture*, pp. 71–100. Washington: Smithsonian Institution.

Dentan, R.K. (1968) *The Semai: A Nonviolent People of Malaya*. New York: Holt, Rinehart and Winston.

Dentan, R.K. (1971) Some Senoi Semai planting techniques. *Economic Botany* **25**: 136–59.

Dentan, R.K., Endicott, K., Gomes, A.G. & Hooker, M.B. (eds.) (1997) *Malaysia and the Original People*. Boston: Allyn and Bacon.

Diffloth, G.F. (1968) Proto-Semai phonology. *Federated Museums Journal* **13**: 65–74.

Dingle, H. (1996) *Migration: The Biology of Life on the Move*. Oxford: Oxford University Press.

Dorit, R.L., Akashi, H. & Gilbert, W. (1995) Absence of polymorphism at the ZFY locus on the human Y chromosome. *Science* **268**: 1183–5.

Douglass, W.A. (1971) Rural exodus in two Spanish Basque villages. *American Anthropologist* **73**: 1100–14.

Douglass, W.A. (1975) *Echalar and Murelaga: Opportunity and Rural Exodus in Two Spanish Basque Villages*. London: C. Hurst and Company.

Douglass, W.A. & Bilbao, J. (1975) *Amerikanuak: Basques in the New World*. Reno: University of Nevada Press.

Dunn, F.L. (1975) *Rain-forest Collectors and Traders: A Study of Resource Utilization in Modern and Ancient Malaya*. Monographs of the Malaysian Branch of the Royal Asiatic Society No. 5. Kuala Lumpur: Royal Asiatic Society.

Dyke, B. (1971) Potential mates in a small human population. *Social Biology* **18**: 28–39.

Dyke, B. (1981) Computer simulation in anthropology. *Annual Review of Anthropology* **10**: 193–207.

Endler, J.A. (1977) *Geographic Variation, Speciation, and Clines*. Princeton: Princeton University Press.

Epperson, B.K. (1990) Spatial autocorrelation of genotypes under directional selection. *Genetics* **124**: 757–771.

Epperson, B.K. (1993) Recent advances in correlation studies of spatial patterns of genetic variation. *Evolutionary Biology* **27**: 95–155.

Epperson, B.K. (1994) Spatial and space-time correlations in systems of subpopulations with stochastic migration. *Theoretical Population Biology* **46**: 160–97.

Evans, I.H.N. (1915) Notes on the Sakai of the Ulu Sungkai in the Batang Padang District of Perak. *Journal of the Federated Malay States Museums* **6**: 85–100.

Ewens, W.J., Brockwell, P.J., Gani, J.M. & Resnick, S.I. (1987) Minimum viable population size in the presence of catastrophes. In: M.E. Soule (ed.). *Viable Populations for Conservation*, pp. 59–68. Cambridge: Cambridge University Press.

Excoffier, L. (1990) Evolution of human mitochondrial DNA: Evidence for departure from a pure neutral model of populations at equilibrium. *Journal of Molecular Evolution* **30**: 125–39.

Felsenstein, J. (1975) A pain in the torus: some difficulties with models of isolation by distance. *American Naturalist* **109**: 359–68.

Felsenstein, J. (1982) How can we infer geography and history from gene frequencies? *Journal of Theoretical Biology* **96**: 9–20.

Fisher, R.A. (1937) The wave of advance of an advantageous allele. *Annals of Eugenics* **7**: 355–69.

Fix, A.G. (1974) Neighbourhood knowledge and marriage distance: the Semai case. *Annals of Human Genetics, London* **37**: 327–32.

Fix, A.G. (1975) Fission–fusion and lineal effect: aspects of the population structure of the Semai Senoi of Malaysia. *American Journal of Physical Anthropology* **43**: 295–302.

Fix, A.G. (1977) *The Demography of the Semai Senoi*. Anthropological Papers No. 62. Ann Arbor: University of Michigan Museum of Anthropology.

Fix, A.G. (1978) The role of kin-structured migration in genetic microdifferentiation. *Annals of Human Genetics, London* **41**: 329–39.

220   *References*

Fix, A.G. (1979) Anthropological genetics of small populations. *Annual Review of Anthropology* **8**: 207–30.

Fix, A.G. (1981) Kin-structured migration and the rate of advance of an advantageous gene. *American Journal of Physical Anthropology* **55**: 433–42.

Fix, A.G. (1982a) Endogamy in settlement populations of Semai Senoi: potential mate pool analysis and simulation. *Social Biology* **28**: 62–74.

Fix, A.G. (1982b) Genetic structure of the Semai. In: M.H. Crawford & J.H. Mielke (eds.), *Current Developments in Anthropological Genetics. Ecology and Population Structure*, pp. 179–204. New York: Plenum.

Fix, A.G. (1984) Kin groups and trait groups: population structure and epidemic disease selection. *American Journal of Physical Anthropology* **65**: 201–12.

Fix, A.G. (1991) Changing sex ratio of mortality in the Semai Senoi, 1969–1987. *Human Biology* **63**: 211–20.

Fix, A.G. (1993) Kin-structured migration and isolation by distance. *Human Biology* **65**: 193–210.

Fix, A.G. (1994) Detecting clinal and balanced selection using spatial autocorrelation analysis under kin-structured migration. *American Journal of Physical Anthropology* **95**: 385–97.

Fix, A.G. (1995) Malayan paleosociology: implications for patterns of genetic variation among the Orang Asli. *American Anthropologist* **97**: 313–23.

Fix, A.G. (1996) Gene frequency clines in Europe: demic diffusion or natural selection? *Journal of the Royal Anthropological Institute* **2 (N.S.)**: 625–43.

Fix, A.G. (1997) Gene frequency clines produced by kin-structured founder effects. *Human Biology* **69**: 663–73.

Fix, A.G. & Lie-Injo, L.E. (1975) Genetic microdifferentiation in the Semai Senoi of Malaysia. *American Journal of Physical Anthropology* **43**: 47–55.

Flatz, G. (1967) Hemoglobin E: distribution and population dynamics. *Humangenetik* **3**: 189–234.

Foley, R. (1987) *Another Unique Species: Patterns in Human Evolutionary Ecology.* New York: Wiley.

Frayer, D.W., Wolpoff, M.H., Thorne, A.G., Smith, F.H. & Pope, G.G. (1993) Theories of modern human origins: the paleontological test. *American Anthropologist* **95**: 14–50.

Friedlaender, J.S. (1975) *Patterns of Human Variation.* Cambridge, MA: Harvard University Press.

Gamble, C. (1994) *Timewalkers: The Prehistory of Global Colonization.* Cambridge, MA: Harvard University Press.

Gibbons, A. (1997) Y chromosome shows that Adam was an African. *Science* **278**: 804–5.

Gladwin, C.H. (1989) *Ethnographic Decision Tree Modeling.* Newbury Park: Sage.

Goody, J. (1976) *Production and Reproduction: A Comparative Study of the Domestic Domain.* Cambridge: Cambridge University Press.

Gould, H.A. (1960) The microdemography of marriages in a North Indian area. *Southwestern Journal of Anthropology* **16**: 476–91.

Gould, S.J. (1985) *The Flamingo's Smile: Reflections in Natural History.* New York: W.W. Norton.

Gould, S.J. (1989) *Wonderful Life: The Burgess Shale and the Nature of History.* New York: W.W. Norton.

Gregg, S.A. (1988) *Foragers and Farmers: Population Interaction and Agricultural Expansion in Prehistoric Europe.* Chicago: University of Chicago Press.

Grigg, D.B. (1977) E.G. Ravenstein and the 'laws of migration'. *Journal of Historical Geography* **3**: 41–54.

Hägerstrand, T. (1967) *Innovation Diffusion as a Spatial Process.* Chicago: University of Chicago Press.

Haggett, P. (1966) *Locational Analysis in Human Geography.* New York: St. Martin's Press.

Haldane, J.B.S. (1949) Disease and evolution. *Ricerca Scientifica (Suppl.)* **19**: 3–10.

Hamilton, A. (1982) Descended from father, belonging to country: rights to land in the Australian Western Desert. In: E. Leacock & R.B. Lee (eds.). *Politics and History in Band Societies*, pp. 85–108. Cambridge: Cambridge University Press.

Hamilton, W.D. (1964) The genetical evolution of social behavior I and II. *Journal of Theoretical Biology* **7**: 1–52.

Hammel, E.A., McDaniel, C.K. & Wachter, K.W. (1979) Demographic consequences of incest tabus: a microsimulation analysis. *Science* **205**: 972–7.

Hammel, E.A., McDaniel, C.K. & Wachter, K.W. (1980) Vice in the Villefranchian: a microsimulation analysis of incest prohibitions. In: B. Dyke & W.T. Morrill (eds.). *Genealogical Demography*, pp. 209–34. New York: Academic Press.

Hammer, M.F. & Zegura, S.L. (1996) The role of the Y chromosome in human evolutionary studies. *Evolutionary Anthropology* **5**: 116–34.

Hanihara, T. (1993) Population prehistory of East Asia and the Pacific as viewed from craniofacial morphology: the basic populations in East Asia, VII. *American Journal of Physical Anthropology* **91**: 173–87.

Hanski, I.A. & Gilpin, M.E. (eds.) (1997) *Metapopulation Biology: Ecology, Genetics, and Evolution.* San Diego: Academic Press.

Harding, R.M., Rosing, F.W., & Sokal, R.R. (1990) Cranial measurements do not support Neolithization of Europe by demic expansion. *Homo* **40**: 45–58.

Harpending, H. & Jenkins, T. (1974) !Kung population structure. In: J.F. Crow & C. Denniston (eds.). *Genetic Distance*, pp. 137–65. New York: Plenum.

Harpending, H., Relethford, J. & Sherry, S.T. (1996) Methods and models for understanding human diversity. In: A.J. Boyce & C.G.N. Mascie-Taylor (eds.). *Molecular Biology and Human Diversity*, pp. 283–99. Cambridge: Cambridge University Press.

Harpending, H., Sherry, S., Rogers, A. & Stoneking, M. (1993) The genetic structure of ancient human populations. *Current Anthropology* **34**: 483–96.

Harpending, H. & Wandsnider, L. (1982) Population structures of Ghanzi and Ngamiland !Kung. In: M.H. Crawford & J.H. Mielke (eds.). *Current Developments in Anthropological Genetics: Ecology and Population Structure*, pp. 29–50. New York: Plenum.

Harpending, H. & Ward, R. (1982) Chemical systematics and human populations. In: M. Nitecki (ed.). *Biochemical Aspects of Evolutionary Biology*, pp. 213–56. Chicago: University of Chicago Press.

Harrison, G.A. (1995) *The Human Biology of the English Village.* Oxford: Oxford University Press.

Harrison, G.A. & Boyce, A.J. (1972a) Migration, exchange, and the genetic structure of human populations. In: G.A. Harrison & A.J. Boyce (eds.). *Structure of Human Populations*, pp. 128–45. Oxford: Clarendon.

Harrison, G.A. & Boyce, A.J. (eds.) (1972b) *The Structure of Human Populations*. Oxford: Clarendon Press.

Headland, T.N. & Reid, L.A. (1989) Hunter–gatherers and their neighbors from prehistory to the present. *Current Anthropology* **30**: 43–51.

Helm, J. (1968) The nature of Dogrib socioterritorial groups. In: R.B. Lee & I. DeVore (eds.). *Man the Hunter*, pp. 118–25. Chicago: Aldine.

Hewlett, B., Koppel, J.M.H. van de & Cavalli-Sforza, L.L. (1982) Exploration ranges of Aka Pygmies of the Central African Republic. *Man* **17**: 418–30.

Hiatt, L.R. (1996) *Arguments About Aborigines: Australia and the Evolution of Social Anthropology*. Cambridge: Cambridge University Press.

Hill, A.V.S. (1991) HLA associations with malaria in Africa: some implications for MHC evolution. In: J. Klein & D. Klein (eds.). *Molecular Evolution of the Major Histocompatibility Complex*, pp. 403–20. Berlin: Springer.

Hiorns, R.W., Harrison, G.A., Boyce, A.J. & Kuchemann, C.F. (1969) A mathematical analysis of the effects of movement on the relatedness between populations. *Annals of Human Genetics, London* **32**: 237–50.

Howell, F.C. (1968) The use of ethnography in reconstructing the past. In: R.B. Lee & I. DeVore (eds.). *Man the Hunter*, pp. 287–8. Chicago: Aldine.

Howell, N. (1979) *Demography of the Dobe !Kung*. New York: Academic Press.

Hurd, J.P. (1983) Kin relatedness and church fissioning among the 'Nebraska' Amish of Pennsylvania. *Social Biology* **30**: 59–66.

Imaizumi, Y., Morton, N.E. & Harris, D.E. (1970) Isolation by distance in artificial populations. *Genetics* **66**: 569–82.

Imanishi, T. & Gojobori, T. (1992) Patterns of nucleotide substitutions inferred from the phylogenies of the class I major histocompatibility complex genes. *Journal of Molecular Evolution* **35**: 196–204.

Isaac, G.L. (1978) The food-sharing behavior of protohuman hominids. *Scientific American* **238**: 90–108.

Jackes, M., Lubell, D. & Meiklehohn, C. (1997a) On physical anthropological aspects of the Mesolithic–Neolithic transition in the Iberian Peninsula. *Current Anthropology* **38**: 839–46.

Jackes, M., Lubell, D. & Meiklejohn, C. (1997b) Healthy but mortal: human biology and the first farmers of western Europe. *Antiquity* **71**: 639–58.

Jackson, J.E. (1976) Vaupes marriage: a network system in the northwest Amazon. In: C.A. Smith (ed.). *Regional Analysis*, vol. 2. *Social Systems*, pp. 65–93. New York: Academic Press.

Jackson, J.E. (1983) *The Fish People: Linguistic Exogamy and Tukanoan Identity in Northwest Amazon*. Cambridge: Cambridge University Press.

Jacquard, A. (1975) Inbreeding: one word, several meanings. *Theoretical Population Biology* **7**: 338–63.

Johnson, A.W. & Earle, T. (1987) *The Evolution of Human Societies: From Foraging Group to Agrarian State*. Stanford: Stanford University Press.

Jorde, L. (1980) The genetic structure of subdivided human populations: a review. In: J.H. Mielke & M.H. Crawford (ed.). *Current Developments in Anthropological Genetics*, pp. 135–208. New York: Plenum.

Jorde, L.B., Workman, P.L. & Eriksson, A.W. (1982) Genetic microevolution in the Åland Islands, Finland. In: M.H. Crawford & J.H. Mielke (eds.). *Current Developments in Anthropological Genetics: Ecology and Population Structure*, pp. 333–66. New York: Plenum.

Kimura, M. & Maruyama, T. (1971) Pattern of neutral polymorphism in a geographically structured population. *Genetics Research, Cambridge* **18**: 125–31.

Kimura, M. & Weiss, G.H. (1964) The stepping stone model of population structure and the decrease of genetic correlation with distance. *Genetics* **49**: 561–76.

Klein, J. (1986) *Natural History of the Major Histocompatibility Complex.* New York: John Wiley.

Korey, K.A. (1996) Some implications of a simulation approach to reconstructing human origins from genetic data. *American Journal of Physical Anthropology* (Supp. 22): 141.

Kunstadter, P., Buhler, R., Stephen, F. & Westoff, C.F. (1963) Demographic variability and preferential marriage patterns. *American Journal of Physical Anthropology* **21**: 511–19.

Kurland, J.A. & Beckerman, S.J. (1985) Optimal foraging and hominid evolution: Labor and reciprocity. *American Anthropologist* **87**: 73–93.

Lahr, M.M. (1996) *The Evolution of Modern Human Diversity.* Cambridge: Cambridge University Press.

Lalueza Fox, C., Gonzalez Martin, A. & Vives Civit, S. (1996) Cranial variation in the Iberian Peninsula and the Balearic Islands: inferences about the history of the population. *American Journal of Physical Anthropology* **99**: 413–28.

Lasker, G.W. & Crews, D.E. (1996) Behavioral influences on the evolution of human genetic diversity. *Molecular Phylogenetics and Evolution* **5**: 232–40.

Lattimore, O. (1951) *Inner Asian Frontiers of China.* New York: American Geographical Society.

Lee, E.S. (1966) A theory of migration. *Demography* **3**: 47–57.

Lee, R.B. (1980) Lactation, ovulation, infanticide, and women's work: A study of hunter–gatherer population regulation. In: M.N. Cohen, R.S. Malpass & H.G. Klein (eds.). *Biosocial Mechanisms of Population Regulation,* pp. 321–48. New Haven: Yale University Press.

Lee, R.B. & DeVore, I. (eds.) (1968) *Man the Hunter.* Chicago: Aldine.

Lee, R.B. & DeVore, I. (eds.) (1976) *Kalahari Hunter–gatherers: Studies of the !Kung San and Their Neighbors.* Cambridge: Harvard University Press.

Leslie, P.W. (1980) Internal migration and genetic differentiation in St. Barthelemy, French West Indies. In: B. Dyke & W.T. Morrill (eds.). *Genealogical Demography,* pp. 167–77. New York: Academic Press.

Leslie, P.W., Dyke, B. & Morrill, W.T. (1980) Celibacy, emigration, and genetic structure in small populations. *Human Biology* **52**: 115–30.

Levin, B.R. & Kilmer, W.L. (1974) Interdemic selection and the evolution of altruism: a computer simulation study. *Evolution* **28**: 527–45.

Levin, D.A. & Fix, A.G. (1989) A model of kin-migration in plants. *Theoretical and Applied Genetics* **77**: 332–6.

Levine, M.H. (1963) Basque isolation: fact or problem? In: V.E. Garfield (ed.). *American Ethnological Society, Proceedings,* pp. 20–31. Seattle: University of Washington Press.

Levins, R. (1966) The strategy of model building in population biology. *American Scientist* **54**: 421–31.

Levins, R. (1970) Extinction. *Some Mathematical Questions in Biology* I, pp. 75–108. Providence: American Mathematical Society.

224   *References*

Lewis, G.J. (1982) *Human Migration: A Geographical Perspective.* London: Croom Helm.
Lewontin, R.C. (1972) The apportionment of human diversity. *Evolutionary Biology* **6**: 381–98.
Lewontin, R.C. (1974) *The Genetic Basis of Evolutionary Change.* New York: Columbia University Press.
Lewontin, R.C. & Krakauer, J. (1973) Distribution of gene frequency as a test of the theory of the selective neutrality of polymorphisms. *Genetics* **74**: 175–95.
Lie-Injo, L.E. (1976) Genetic relationships of several aboriginal groups in South East Asia. In: R.L. Kirk & A.G. Thorne (eds.). *The Origin of the Australians,* pp. 277–306. Canberra: Australian Institute of Aboriginal Studies.
Little, M. & Leslie, P.W. (1993) Migration. In: G.W. Lasker & C. Mascie-Taylor (eds.). *Research Strategies in Human Biology,* pp. 62–91. Cambridge: Cambridge University Press.
Livingstone, F.B. (1958) Anthropological implications of sickle cell gene distribution in West Africa. *American Anthropologist* **60**: 533–62.
Livingstone, F.B. (1962) Population genetics and population ecology. *American Anthropologist* **64**: 44–52.
Livingstone, F.B. (1967) *Abnormal Hemoglobins in Human Populations.* Chicago: Aldine.
Livingstone, F.B. (1969) Gene frequency clines of the β hemoglobin locus in various human populations and their simulation by models involving differential selection. *Human Biology* **41**: 223–36.
Livingstone, F.B. (1976) Hemoglobin history in West Africa. *Human Biology* **48**: 487–500.
Livingstone, F.B. (1985) *Frequencies of Hemoglobin Variants: Thalassemia, the Glucose–6-Phosphate Dehydrogenase Deficiency, G6PD Variants, and Ovalocytosis in Human Populations.* New York: Oxford University Press.
Livingstone, F.B. (1989) Simulation of the diffusion of the B-globin variants in the Old World. *Human Biology* **61**: 267–309.
Livingstone, F.B. (1991) Phylogenies and the forces of evolution. *American Journal of Human Biology* **3**: 83–9.
Lovejoy, C.O. (1981) The origin of man. *Science* **211**: 341–50.
Lowie, R.L. (1937) *The History of Ethnological Theory.* New York: Farrar and Rinehart.
MacCluer, J.W. (1974) Avoidance of incest: genetic and demographic consequences. In: B. Dyke & J.W. MacCluer (eds.). *Computer Simulation in Human Population Studies,* pp. 197–220. New York: Academic Press.
MacCluer, J.W. & Dyke, B. (1976) On the minimum size of endogamous populations. *Social Biology* **23**: 1–12.
Madsen, D.B. & Rhode, D. (eds.) (1994) *Across the West: Human Population Movement and the Expansion of the Numa.* Salt Lake City: University of Utah Press.
Malécot, G. (1950) Quelques schemas probabilistes sur la variabilite des populations naturelles. *Annales de l'Université de Lyon Science, Sect. A* **13**: 37–60.
Malécot, G. (1955) Decrease of relationship in distance. *Cold Spring Harbor Symposium on Quantitative Biology* **20**: 52–3.
Malécot, G. (1973) Isolation by distance. In: N.E. Morton (ed.). *Genetic Structure of*

Populations, pp. 72–5. Honolulu: University of Hawaii Press.

Mallory, J.P. (1989) *In Search of the Indo-Europeans: Language, Archaeology and Myth*. London: Thames and Hudson.

Manderscheid, E., Brannan, J. & Rogers, A.R. (1994) Is migration kin structured? *Human Biology* **66**: 49–58.

Mange, A.P. (1964) Growth and inbreeding of a human isolate. *Human Biology* **36**: 104–33.

Marshall, L. (1976) *The !Kung of Nyae Nyae*. Cambridge, MA: Harvard University Press.

Martin, R. (1905) *Die Inlandstamme der Malayischen Halbinsel*. Jena: Gustav Fischer.

Mascie-Taylor, C.G.N. & Boyce, A.J. (eds.). (1988) *Human Mating Patterns*. Cambridge: Cambridge University Press.

Maynard Smith, J. (1976) Group selection. *Quarterly Review of Biology* **51**: 277–83.

Mayr, E. (1963) *Animal Species and Evolution*. Cambridge, MA: Belknap.

McCullough, J.M. & Barton, E.Y. (1991) Relatedness and kin-stuctured migration in a founding population: Plymouth. *Human Biology* **63**: 355–66.

Meiklejohn, C., Schentag, C., Venema, A. & Key, P. (1984) Socioeconomic change and patterns of pathology and variation in the Mesolithic and Neolithic of Western Europe: some suggestions. In: M.N. Cohen & G.J. Armelagos (eds.). *Paleopathology at the Origins of Agriculture*, pp. 75–100. Orlando: Academic Press.

Mellars, P. & Stringer, C. (eds.) (1989) *The Human Revolution: Behavioral and Biological Perpectives on the Origins of Modern Humans*. Princeton: Princeton University Press.

Melton, T., Peterson, R., Redd, A.J., Saha, N., Sofro, A.S.M., Martinson, J. & Stoneking, M. (1995) Polynesian genetic affinities with southeast Asian populations as identified by mtDNA analysis. *American Journal of Human Genetics* **57**: 403–14.

Menozzi, P., Piazza, A. & Cavalli-Sforza, L. (1978) Synthetic maps of human gene frequencies in Europeans. *Science* **201**: 786–92.

Merrell, D.J. (1981) *Ecological Genetics*. Minneapolis: University of Minnesota Press.

Michod, R.E. & Abugov, R. (1980) Adaptive topography in family-structured models of kin selection. *Science* **210**: 667–9.

Mielke, J.H., Workman, P.C., Fellman, J. & Eriksson, A.W. (1976) Population structure of the Åland Islands, Finland. *Advances in Human Genetics* **6**: 241–321.

Miller, L.H., Mason, S.J., Dvorak, J.A., McGinniss, M.H. & Rothman, I.K. (1975) Erythrocyte receptors for (*Plasmodium knowlesi*) malaria: Duffy blood group determinants. *Science* **189**: 561–3.

Milner, A.C. (1978) A note on 'the Rawa'. *Journal of the Malaysian Branch of the Royal Asiatic Society* **51**: 143–8.

Moore, J.H. (1994) Putting anthropology back together again: the ethnogenetic critique of cladistic theory. *American Anthropologist* **96**: 925–48.

Morgan, K. (1974) Computer simulation of incest prohibition and clan proscription rules in closed, finite populations. In: B. Dyke & J.W. MacCluer (eds.). *Computer Simulation in Human Population Studies*, pp. 15–42. New York: Academic.

Morgan, L.H. (1877, reprinted 1963) *Ancient Society*. Cleveland: World Publishing Co.

Morrill, R.L. (1965) *Migration and the Spread and Growth of Urban Settlement*. Lund Studies in Geography, Series B. No.26.

Morrill, R.L. & Pitts, F.R. (1967) Marriage, migration, and the mean information field: a study in uniqueness and generality. *Annals of the Association of American Geographers* **57**: 401–22.

Morton, N.E. (1972) The future of human population genetics. *Progress in Medical Genetics* **8**: 103–24.

Morton, N.E. (ed.) (1973) *Genetic Structure of Populations*. Honolulu: University of Hawaii Press.

Morton, N.E. (1977) Isolation by distance in human populations. *Annals of Human Genetics, London* **40**: 361–5.

Nagylaki, T. (1979) The island model with stochastic migration. *Genetics* **91**: 163–76.

Neel, J.V. (1967) The genetic structure of primitive human populations. *Japanese Journal of Human Genetics* **12**: 1–16.

Neel, J.V. (1970) Lessons from a 'primitive' people. *Science* **170**: 815–21.

Neel, J.V. (1978) The population structure of an Amerindian tribe, the Yanomama. *Annual Review of Genetics* **12**: 365–413.

Neel, J.V. (1984) The 'real' human populations. In: A. Chakravarti (ed.). *Human Population Genetics: The Pittsburgh Symposium*, pp. 249–73. New York: Van Nostrand Reinhold.

Neel, J.V., Centerwall, W.R., Chagnon, N.A. & Casey, H.O. (1970) Notes on the effect of measles and measle vaccine in a virgin-soil population of South American Indians. *American Journal of Epidemiology* **91**: 419–29.

Neel, J.V. & Salzano, F.M. (1967) Further studies on the Xavante Indians. X. Some hypotheses-generalizations resulting from these studies. *American Journal of Human Genetics* **19**: 554–74.

Neel, J.V. & Weiss, K.M. (1975) The genetic structure of a tribal population, the Yanomama Indians. XII. Biodemographic studies. *American Journal of Physical Anthropology* **42**: 25–51.

Nei, M. & Hughes, A.A.L. (1991) Polymorphism and evolution of the major histocompatibility complex loci in mammals. In: R.K. Selander, A.G. Clark & T.S. Whittam (eds.). *Evolution at the Molecular Level*, pp. 222–47. Sunderland: Sinauer.

Nei, M. & Roychoudhury, A.K. (1974) Genetic variation within and between the three major races of man. *American Journal of Human Genetics* **26**: 421–43.

Newell, C. (1988) *Methods and Models in Demography*. New York: Guilford Press.

Nitecki, M.H. & Nitecki, D.V. (eds.). (1994) *Origins of Anatomically Modern Humans*. New York: Plenum Press.

Noone, H.D. (1936) Report on the settlements and welfare of the Ple-Temiar Senoi of the Perak-Kelantan watershed. *Journal of the Federated Malay States Museums* **19**: 1–85.

Oden, N. (1984) Assessing the significance of a spatial correlogram. *Geographical Analysis* **16**: 1–16.

Olsen, C.L. (1987) The demography of colony fission from 1878–1970 among the Hutterites of North America. *American Anthropologist* **89**: 823–37.

Olsson, G. (1965) Distance and human interaction: a migration study. *Geografiska Annaler* **47**: 3–43.

Pianka, E.R. (1970) On 'r' and 'K' selection. *American Naturalist* **104**: 592–7.

Piazza, A., Menozzi, P. & Cavalli-Sforza, L.L. (1980) The HLA-A,B gene frequencies in the world: migration or selection? *Human Immunology* **4**: 297–304.

Polunin, I. (1953) The medical natural history of the Malayan Aborigines. *Medical Journal of Malaya* **8**: 62–174.

Polunin, I. & Sneath, P.H.A. (1953) Studies of blood groups in South-east Asia. *Journal of the Royal Anthropological Institute* **83**: 215–51.

Potts, W.K. & Wakeland, E.K. (1990) Evolution of diversity at the major histocompatibility complex. *Trends in Evolution and Ecology* **5**: 181–7.

Radcliffe-Brown, A.R. (1930) The social organization of Australian tribes. *Oceania* **1**: 34–65.

Ravenstein, E.G. (1885) The laws of migration. *Journal of the Statistical Society* **48**: 167–227.

Relethford, J.H. (1995) Genetics and modern human origins. *Evolutionary Anthropology* **4**: 53–63.

Relethford, J.H. (1997) Mutation rate and excess African heterozygosity. *Human Biology* **69**: 785–92.

Relethford, J.H. & Harpending, H.C. (1994) Craniometric variation, genetic theory, and modern human origins. *American Journal of Physical Anthropology* **95**: 249–70.

Rendine, S., Piazza, A. & Cavalli-Sforza, L.L. (1986) Simulation and separation by principal components of multiple demic expansions in Europe. *American Naturalist* **128**: 681–706.

Richards, M., Corte-Real, H., Forster, P., Macaulay, V., Wilkinson-Herbots, H., Demaine, A., Papiha, S., Hedges, R., Bandelt, H.-J. & Sykes, B. (1996) Paleolithic and Neolithic lineages in the European mitochondrial gene pool. *American Journal of Human Genetics* **59**: 185–203.

Robarchek, C.A. (1979) Conflict, emotion, and abreaction: resolution of conflict among the Semai Senoi. *Ethnos* **7**: 104–23.

Roberts, D.F. (1967) The development of inbreeding in an island population. *Ciência e Cultura* **19**: 78–84.

Roberts, D.F. (1968) Genetic effects of population size reduction. *Nature* **220**: 1084–8.

Roberts, L. (1992) How to sample the world's genetic diversity. *Science* **257**: 1204–5.

Robinson, V. (ed.) (1996) *Geography and Migration*. The International Library of Studies on Migration. Cheltenham: Elgar Reference Collection.

Rogers, A.R. (1987) A model of kin-structured migration. *Evolution* **41**: 417–26.

Rogers, A.R. (1988) Three components of genetic drift in subdivided populations. *American Journal of Physical Anthropology* **77**: 435–50.

Rogers, A.R. (1995) Genetic evidence for a Pleistocene population explosion. *Evolution* **49**: 608–15.

Rogers, A.R. & Eriksson, A.W. (1988) Statistical analysis of the migration component of genetic drift. *American Journal of Physical Anthropology* **77**: 451–8.

Rogers, A.R. & Harpending, H. (1986) Migration and genetic drift in human populations. *Evolution* **40**: 1312–27.

Rogers, A.R. & Harpending, H. (1992) Population growth makes waves in the distribution of pairwise genetic differences. *Molecular Biology and Evolution* **9**: 552–69.

Rogers, A.R. & Jorde, L.B. (1987) The effect of non-random migration on genetic differences between populations. *Annals of Human Genetics* **51**: 169–76.

Rogers, A.R. & Jorde, L.B. (1995) Genetic evidence on modern human origins. *Human Biology* **67**: 1–36.

Rogers, A.R. & Jorde, L.B. (1996) Ascertainment bias in estimates of average heterozygosity. *American Journal of Human Genetics* **58**: 1033–41.

Saha, N., Mak, J.W., Tay, J.S.H., Liu, Y., Tan, J.A.M.A., Low, P.S. & Singh, M. (1995) Population genetic studies among the Orang Asli (Semai Senoi) of Malaysia: Malayan Aborigines. *Human Biology* **67**: 37–57.

Sahlins, M.D. (1968) *Tribesmen*. Englewood Cliffs: Prentice Hall.

Salzano, R.M. (ed.) (1975) *The Role of Natural Selection in Human Evolution*. New York: American Elsevier/North Holland.

Schebesta, P. (1952) *Die Negrito Asiens. Band I*. Studia Instituti Anthropos **6**. Wein-Modling: St. Gabriel.

Schmidt, W. (1939) *The Culture Historical Method of Ethnology*. New York: Fortuny's.

Serjeantson, S., Bryson, K., Amato, D. & Babona, D. (1977) Malaria and hereditary ovalocytosis. *Human Genetics* **37**: 161–7.

Serjeantson, S.W. & Hill, A.V.S. (1989) Colonization of the Pacific: the genetic evidence. In: A.V.S. Hill & S.W. Serjeantson (eds.) *Colonization of the Pacific: A Genetic Trail*, pp. 286–94. Oxford: Clarendon.

Service, E.R. (1962) *Primitive Social Organization: An Evolutionary Perspective*. New York: Random House.

Shields, W.M. (1987) Dispersal and mating systems: investigating their causal connections. In: B.D. Chepko-Sade & Z.T. Halpin (eds.). *Mammalian Dispersal Patterns*, pp. 3–24. Chicago: University of Chicago Press.

Skeat, W.W. & Blagden, C.O. (1906) *The Pagan Races of the Malay Peninsula*. London: MacMillan.

Skinner, G.W. (1964) Marketing and social structure in rural China: part I. *Journal of Asian Studies* **24**: 3–43.

Slatkin, M. (1981) Populational heritability. *Evolution* **35**: 859–71.

Slatkin, M. & Arter, H.E. (1991) Spatial autocorrelation methods in population genetics. *American Naturalist* **138**: 499–517.

Slicher van Bath, B.H. (1963) *The Agrarian History of Western Europe, A.D. 500–1850*. New York: St. Martin's Press.

Smouse, P.E. (1982) Genetic architecture of swidden agricultural tribes from the lowland rain forests of South America. In: M.H. Crawford & J.H. Mielke (eds.). *Current Developments in Anthropological Genetics: Ecology and Population Structure*, pp. 139–78. New York: Plenum.

Smouse, P.E., Vitzthum, V.J. & Neel, J.V. (1981) The impact of random and lineal fission on the genetic divergence of small human groups: a case study among the Yanomama. *Genetics* **98**: 179–91.

Smouse, P.E. & Long, J.C. (1988) A comparative F-statistics analysis of the genetic structure of human populations from lowland South America and Highland New Guinea. In: B.S. Weir (ed.). *Proceedings of the Second International Conference on Quantitative Genetics*, pp. 32–46. Sunderland, MA: Sinauer.

Smouse, P.E. & Wood, J.W. (1987) The genetic demography of the Gainj of Papua New Guinea: functional models of migration and their genetic implications. In: B.D. Chepko-Sade & Z.T. Halpin (eds.). *Mammalian Disperal Patterns: The Effects of Social Structure on Population Genetics*. Chicago: University of Chicago Press.

Sofro, A.D.M. (1986) Ovalocytosis in Indonesia: distribution and its relation to the malaria hypothesis. *Medika* **10**: 954–8.

Sokal, R.R., Harding, R.M. & Oden, N.L. (1989a) Spatial patterns of human gene frequencies in Europe. *American Journal of Physical Anthropology* **80**: 267–94.

Sokal, R.R., Jacquez, G.M. & Wooten, M.C. (1989b) Spatial autocorrelation analysis of migration and selection. *Genetics* **121**: 845–55.

Sokal, R.R. & Menozzi, P. (1982) Spatial autocorrelations of HLA frequencies in Europe support demic diffusion of early farmers. *American Naturalist* **119**: 1–17.

Sokal, R.R., Oden, N.L., Walker, J., Di Giovanni, D. & Thomson, B.A. (1996) Historical population movements in Europe influence genetic relationships in modern samples. *Human Biology* **68**: 873–98.

Sokal, R.R., Oden, N.L. & Wilson, C. (1991) Genetic evidence for the spread of agriculture in Europe by demic diffusion. *Nature* **351**: 143–5.

Sokal, R.R. & Wartenburg, D.E. (1983) A test of spatial autocorrelation analysis using an isolation-by-distance model. *Genetics* **105**: 219–37.

Solheim, W.G. (1980) Searching for the origins of the Orang Asli. *Federated Museums Journal* **25**: 61–76.

Soulé, M.E. (ed.) (1987) *Viable Populations for Conservation*. Cambridge: Cambridge University Press.

Spieth, P.T. (1974) Gene flow and genetic differentiation. *Genetics* **78**: 961–5.

Stearns, S.C. (1992) *The Evolution of Life Histories*. Oxford: Oxford University Press.

Steward, J.H. (1955) *Theory of Culture Change: The Methodology of Multilinear Evolution*. Urbana: University of Illinois Press.

Strauss, D.J. & Orans, M. (1975) Mighty sifts: a critical appraisal of solutions to Galton's problem and a partial solution. *Current Anthropology* **16**: 573–94.

Stringer, C.B. & Andrews, P. (1988) Genetic and fossil evidence for the origin of modern humans. *Science* **239**: 1263–8.

Sved, J.A. & Latter, B.D.H. (1977) Migration and mutation in stochastic models of gene frequency change. *Journal of Mathematical Biology* **5**: 61–73.

Swedlund, A.C. (1972) Observations on the concept of neighbourhood knowledge and the distribution of marriage distances. *Annals of Human Genetics, London* **35**: 327–30.

Swedlund, A.C. (1980) Historical demography: applications in anthropological genetics. In: J.H. Mielke & M.H. Crawford (eds.). *Current Developments in Anthropological Genetics*, vol. 1, *Theory and Methods*. New York: Plenum.

Templeton, A.R. (1992) Human origins and the analysis of mitochondrial DNA sequences. *Science* **255**: 737.

Templeton, A.R. (1993) The 'Eve' hypotheses: a genetic critique and reanalysis. *American Anthropologist* **95**: 51–72.

Tindale, N.B. (1953) Tribal and intertribal marriage among the Australian Aborigines. *Human Biology* **25**: 169–90.

Tonkinson, R. (1978) *The Mardudjara Aborigines: Living the Dream in Australia's Desert*. New York: Holt, Rinehart and Winston.

230   *References*

Tooby, J. & DeVore, I. (1987) The reconstruction of hominid behavioral evolution through strategic modeling. In: W.G. Kinzey (ed.). *The Evolution of Human Behavior: Primate Models*, pp. 183–237. Albany, NY: State University Press of New York.

Turnbull, C.M. (1961) *The Forest People*. New York: Simon and Schuster.

Turnbull, C.M. (1968) The importance of flux in two hunting societies. In: R.B. Lee & I. DeVore (eds.). *Man the Hunter*, pp. 12–32–137. Chicago: Aldine.

Turnbull, C.M. (1986) Survival factors among Mbuti and other hunters of the equatorial African rain forest. In: L.L. Cavalli-Sforza (ed.). *African Pygmies*, pp. 103–23. Orlando: Academic Press.

Turnbull, H.F. (1981) The importance of population structure and birth order specific selection in relation to the maintenance and distribution of the Rhesus blood group polymorphism in human populations. PhD: University of California, Riverside.

Turner, C.G. (1990) Major features of Sundadonty and Sinodonty, including suggestions about East Asian microevolution, population history, and late Pleistocene relationships with Australian Aboriginals. *American Journal of Physical Anthropology* **82**: 295–317.

Tylor, E.B. (1888) On a method of investigating the development of institutions; applied to laws of marriage and descent. *Journal of the Royal Anthropological Institute* **18**: 245–69.

Wade, M.J. (1978) A critical review of the models of group selection. *Quarterly Review of Biology* **53**: 101–14.

Wagener, D.K. (1973) An extension of migration matrix analysis to account for differential immigration from the outside world. *American Journal of Human Genetics* **25**: 47–56.

Wallace, D. & Torroni, A. (1992) American Indian prehistory as written by mtDNA: a review. *Human Biology* **64**: 403–16.

Wallace, D.C. (1997) Mitochondrial DNA in aging and disease. *Scientific American* **277**: 40–7.

Ward, R.H. & Neel, J.V. (1970) Gene frequencies and microdifferentiation among Makiritare Indians. IV. A comparison of a genetic network with ethnohistory and migration matrices; a new index of genetic isolation. *American Journal of Human Genetics* **22**: 538–61.

Ward, R.H. & Neel, J.V. (1976) The genetic structure of a tribal population, the Yanomama Indians. XIV. Clines and their interpretation. *Genetics* **82**: 103–21.

Weiss, K.M. (1988) In search of times past: gene flow and invasion in the generation of human diversity. In: C.G.N. Mascie-Taylor & G. Lasker (eds.). *Biological Aspects of Human Migration*, pp. 130–66. Cambridge: Cambridge University Press.

Weiss, K.M. (1993) *Genetic Variation and Human Disease: Principles and Evolutionary Approaches*. Cambridge: Cambridge University Press.

Weiss, K.M. & Maruyama, T. (1976) Archeology, population genetics and studies of human racial ancestry. *Americal Journal of Physical Anthropology* **44**: 31–49.

White, N.G. (1989) Cultural influences on the biology of Aboriginal people: examples from Arnhem Land. In: L.H. Schmitt, L. Freeman & B. N.W. (eds.). *The Growing Scope of Human Biology*. Proceedings of the Australasian Society for Human Biology, No. 2, pp. 171–8. Canberra: University of Western Australia.

White, N.G. (1995) Inside the gurrnganngara: social processes and demographic genetics in north-east Arnhem Land, Australia. In: A.J. Boyce & V. Reynolds (eds.). *Human Populations: Diversity and Adaptation*, pp. 252–79. Oxford: Oxford University Press.

Wiessner, P. (1982) Risk, reciprocity and social influences on !Kung San economics. In: E. Leacock & R.B. Lee (eds.). *Politics and History in Band Societies*, pp. 61–84. Cambridge: Cambridge University Press.

Williams, B.J. (1974) *A Model of Band Society*. Memoirs of the Society for American Archaeology, No. 29.

Williams, G.C. (1966) *Adaptation and Natural Selection*. Princeton: Princeton University Press.

Williams, N.M. & Hunn, E.S. (eds) (1982) *Resource Managers: North American and Australian Hunter–Gatherers*. Washington: American Association for the Advancement of Science.

Williams-Hunt, P.D.R. (1951) *An Introduction to the Malayan Aborigines*. Kuala Lumpur: The Government Press.

Wilson, D.S. (1975) A theory of group selection. *Proceedings of the National Academy of Sciences, USA* **72**: 143–6.

Wilson, D.S. (1980) *The Natural Selection of Populations and Communities*. Menlo Park: Benjamin/Cummings.

Wise, C.A., Sraml, M., Rubinztein, D.C. & Easteal, S. (1997) Comparative nuclear and mitochondrial genome diversity in humans and chimpanzees. *Molecular Biology and Evolution* **14**: 707–17.

Wobst, H.M. (1978) The archaeo-ethnology of hunter–gatherers or the tyranny of the ethnographic record in archaeology. *American Antiquity* **43**: 303–9.

Wolpoff, M.H. (1989) Multiregional evolution: the fossil alternative to Eden. In: P. Mellars & C. Stringer (eds.). *The Human Revolution: Behavioural and Biological Perpectives on the Origins of Modern Humans*, pp. 62–108. Princeton: Princeton University Press.

Wolpoff, M.H., Thorne, A.G., Smith, F.H., Frayer, D.W. & Pope, G.G. (1994) Multiregional evolution: a world-wide source for modern human populations. In: M.H. Nitecki & D.V. Nitecki (eds.). *Origins of Anatomically Modern Humans*, pp. 175–99. New York: Plenum.

Wood, J.W. (1977) A stability test for migration matrix models of genetic differentiation. *Human Biology* **49**: 309–20.

Wood, J.W. & Smouse, P.E. (1982) A method of analyzing density-dependent vital rates with an application to the Gainj of Papua New Guinea. *American Journal of Physical Anthropology* **58**: 403–11.

Wood, J.W., Smouse, P.E. & Long, J.C. (1985) Sex-specific dispersal patterns in two human populations of highland New Guinea. *American Naturalist* **125**: 747–68.

Workman, P.L., Mielke, J.H. & Nevanlinna, H.R. (1976) The genetic structure of Finland. *American Journal of Physical Anthropology* **44**: 341–68.

Wright, S. (1931) Evolution in mendelian populations. *Genetics* **16**: 97–159.

Wright, S. (1943) Isolation by distance. *Genetics* **28**: 114–38.

Wright, S. (1948) On the roles of directed and random changes in gene frequency in the genetics of populations. *Evolution* **2**: 279–94.

232   *References*

Wright, S. (1951) The genetical structure of populations. *Annals of Eugenics* **15**: 323–54.
Wright, S. (1955) Classification of the factors of evolution. *Cold Spring Harbor Symposium on Quantitative Biology* **20**: 16–24.
Wright, S. (1969) *Evolution and the Genetics of Populations*, vol. 2, *The Theory of Gene Frequencies*. Chicago: University of Chicago Press.
Wynne-Edwards, V.C. (1962) *Animal Dispersion in Relation to Social Behavior*. Edinburgh: Oliver and Boyd.
Yellen, J. & Harpending, H. (1972) Hunter–gatherer populations and archaeological inference. *World Archaeology* **4**: 244–53.
Yellen, J.E. (1977) *Archaeological Approaches to the Present: Models for Reconstructing the Past*. New York: Academic Press.

Index